THE GREAT GREENWASHING

Published by arrangement with Melbourne Books.
Published in Canada in 2024 and the USA in 2024 by House of Anansi Press Inc.
houseofanansi.com

House of Anansi Press is committed to protecting our natural environment. This book is made of material from well-managed FSC®-certified forests, recycled materials, and other controlled sources.

House of Anansi Press is a Global Certified Accessible™ (GCA by Benetech) publisher. The ebook version of this book meets stringent accessibility standards and is available to readers with print disabilities.

28 27 26 25 24   1 2 3 4 5

Library and Archives Canada Cataloguing in Publication

Title: The great greenwashing : how brands, governments,
and influencers are lying to you / John Pabon.
Names: Pabon, John, author.
Description: Includes bibliographical references.
Identifiers: Canadiana (print) 20230548970 | Canadiana (ebook) 20230549004 |
ISBN 9781487012861 (softcover) | ISBN 9781487012878 (EPUB)
Subjects: LCSH: Consumption (Economics)—Moral and ethical aspects. | LCSH: Greenwashing.
Classification: LCC HC79.C6 P33 2024 | DDC 339.47—dc23

House of Anansi Press is grateful for the privilege to work on and create from the Traditional Territory of many Nations, including the Anishinabeg, the Wendat, and the Haudenosaunee, as well as the Treaty Lands of the Mississaugas of the Credit.

With the participation of the Government of Canada
Avec la participation du gouvernement du Canada

We acknowledge for their financial support of our publishing program the Canada Council for the Arts, the Ontario Arts Council, and the Government of Canada.

Printed and bound in Canada

THE GREAT GREENWASHING

How Brands, Governments, and Influencers Are Lying to You

JOHN PABON

ANANSI
INTERNATIONAL

Contents

PART 1: GREENWASHING: A CRASH COURSE   7

1   Did You Hang Up Your Towel?   8
2   How to Spot Greenwashing   16
3   More Than Just Green   30

PART 2: THE CORPORATE SECTOR   39

4   The Good, the Bad, and the Ugly   40
5   The Un-Sustainable   47
6   Those You Can Trust   65
7   Those That Can Change   75
8   Fast Fashion   97
9   Perfect Be Damned   125

PART 3: STATE-SPONSORED GREENWASHING   131

10   Paris, *je t'aime*   132
11   COP Out   136
12   Toeing the Line   156
13   Hyper-National Organisations   180
14   Running in Place   199

PART 4: INFLUENCES   205

15   Hey There, Sausage Fingers   206
16   Activists, Moralists, and NGOs ... Oh My!   210
17   Bright Lights Cast Long Shadows   226
18   You ... Yes, You   250
19   Build Your Army   265

A Future Worth Fighting For   269

Endnotes   274

The Author   294

PART 1

GREENWASHING: A CRASH COURSE

Chapter 1

Did You Hang Up Your Towel?

There you are, staring down the long grocery store aisle stretching out in front of you. As the musak plays through the store's speakers, you think to yourself how much you used to like the Backstreet Boys. Bopping your head ever so slightly, you start down the aisle. The shiny linoleum, freshly polished, reflects the god-awful bright lights from overhead. On either side are shelves and shelves of products in all manner of packaging. It's like a major freeway. The five, six, and seven shelves are bumper to bumper on either side as you stand atop the centre median.

You stop midway down the aisle between the crisps and biscuits to contemplate the aptly labelled section, 'coffee and tea'. Looking down to the bottom shelf, you spot the ulcer-inducing coffee crystals your grandparents swear by. Flanking these are store-brand products at rock-bottom prices. Scanning the top shelf, a picked-through array of extras—creamers, sugars, disposable cups—sit dishevelled. Right in front of you, just at eye level, is what you came looking for. Boxes and boxes of teas from all over the world. Assam from India. Jasmine from China. English breakfast from, well, England.

You're really spoilt for choice. There's the stock-standard yellow box, more economical for a family since it's loaded with nearly one hundred teabags. The uber-fancy teas usually have some sort of bowel-cleansing properties. Smack bang in the middle is a wide range of excellent teas all priced competitively to one another. You're no tea expert, so how is one to choose?

Standing there, you mull over the options for what's probably way too long. Then you spot a box claiming to be fully organic. It even has a little label certifying it as sustainably sourced. On the back of the box you read about how this brand is a social enterprise, helping out the people who make the tea. Plus, it's no more expensive than anything else on the shelf. Why wouldn't you buy this brand? It seems a no-brainer.

After paying, you grab your calico bag (the millionth you've bought since you keep forgetting the others at home) and walk back to the car, confident in having done a good deed for the day. You've spoken with your wallet, positively impacting the world. Not only that, but you've found a trustworthy product. Next time you head to the store, you'll know exactly what to buy. You'll probably even talk to your friends, family, and colleagues about it. This deal is just too good not to share. Once you get home, you can rest easy. Maybe this whole green living thing isn't that difficult after all.

And just like that, you've fallen victim to greenwashing.

It happens to the best of us. That's because companies looking to make a quick buck are inundating us with heaps of slick messaging. But words only tell part of the story. In this book, you'll learn how to decipher this messaging, especially anything about saving the Earth. We're going to take you from being a passive observer to someone who can push for change. You'll have to learn to read between the lines to get there. Like a good journalist, it'll be your job to get to the true heart of a matter. Now's not the time to take prisoners or pussyfoot around the issues. In this game, you've got to be brave enough to name and shame.

That's because, for too long, many of the companies, governments, and individuals claiming to do the most are nowhere near walking their talk. They've put up beautiful window dressing to distract from all the horrible stuff happening inside their houses. Behind the scenes, these groups' actions hold back the very progress they claim to support. We've been treated like fools, falling into the trap of believing their lies and refusing to question their sincerity. Greenwashing is the result.

We'll see how these groups fail to meet their stated environmental commitments. A good starting point when learning to read between the lines, though, is the corporate sector. Corporate business tends to be the most vocal in supporting a more sustainable future. Of course, they actualise this to varying degrees. I'm in no way trying to say all corporate actors are bad. Later on, I'll explain how the private sector is in the best position to have the most positive impact. For now, I'm speaking directly about those companies engaged in the pesky little practice of *greenwashing*.

Some of you may have heard this term before. Many of you would have unknowingly fallen victim to it. That's why it's vital to equip you with the knowledge to help question everything corporations claim. It doesn't matter if it's on a product's packaging, advertising, or talking face-to-face with corporate executives. Critical thinking has become an underused skill in today's society, where we are constantly walking on eggshells to avoid offending anyone. I'll tell you straight up: the world is burning, so there's no time to cry over spilt milk. We must hold those in power accountable, regardless of whether they sit on Capitol Hill or in the corner office.

What exactly is greenwashing? *Investopedia* does an excellent job at defining the term when it says, '... greenwashing is conveying a false impression that a company or its products are more environmentally sound than they really are.'[1] It's a play on 'whitewashing'; a term used when a company tries to put a good spin on something bad. The same holds for greenwashing. Companies engaging in greenwashing are just trying to put a good spin on their poor behaviour. Perhaps it's on the environmental sustainability of a product or service. Maybe it goes as far as to position a company as eco-friendly when, in reality, they aren't. No matter its use, greenwashing at its core is a lie.

The term itself was coined back in the mid-1980s by environmentalist Jay Westerveld. One has to transport themselves back to this period to truly understand the word's origin. Synonymous with greed, indulgence, and questionable fashion choices, the 1980s were a time of big business getting bigger. In the developed world, it was the era of pure capitalism. Australian social scientist Alex Carey once posited, '... the growth of democracy; the growth of corporate power; and the growth of corporate propaganda as a means of protecting corporate power against democracy' are the most significant political developments of the twentieth century.[2] The 1980s were the decade where all three of these things took hold.

Until the end of that decade, small businesses still employed about a third of American workers. A New York Times analysis notes small businesses accounted for many more jobs overall than large corporations.[3] Then, the mergers began. During the late 1980s and the early 1990s, a large number of business conglomerates began to form. From KKR purchasing Nabisco for a cool US$68 billion in today's dollars to Vodafone buying out Mannesman for US$304 billion, corporations were becoming more prominent and more powerful.[4]

This growth hasn't stopped, either. Think about what your high street shopping strip looked like a couple of decades ago. Remember the old mom-and-pop shop where you could buy a hammer, electrical tape, and a pack of gum? Now, it's probably a big-box home improvement store. How about the local grocery store being replaced by a Coles, Walmart, or Wholefoods? We continue to see consolidation across industries. To put things into perspective, just ten brands own most of today's consumer packaged goods, and only six companies control most media.[5] Although we might think we have a choice when it comes to spending our money, that choice is an illusion.

Perhaps it was this increasingly complicated web of conglomeration, confusing bureaucracy, or sleight of hand. Over the two decades between 1980 and 2000, corporations seemed to think they were invincible. Their new levels of leverage, coupled with consumers' inability to vote with their wallets, resulted in greed and hubris of biblical proportions. A personal favourite example is the 1998 Cendant Corporation accounting incident; one of the largest financial scandals of the 1990s.[6] Cendant reported US$500 million in false profits, costing shareholders a whopping US$19 billion. A corporation imagining they could get away with this as if it were a simple accounting error shows how indomitable they saw themselves.

This corporate megalomania, combined with a lack of access to information by the public, gave *carte blanche* to twisting facts.

But the 1980s were just where greenwashing took off. The practice is much older. One of the earliest examples of corporate greenwashing is something we take for granted today. Think about the last time you stayed in a hotel somewhere. What did you do with your towels? If you're like most people, you hung them up to reuse. Why? Because that little card on the bathroom sink reminded you. These cards, originating in the 1960s, tell us to save the Earth by reusing our towels. They often have statistics about how much water we waste in the shower or how many people go without clean drinking water each year. You're encouraged to do your part by keeping towels out of the wash, thus saving millions of gallons of water.

On the surface, it seems like such a simple thing: individual action making a substantial positive impact. In many ways, this is undoubtedly true. Laundry accounts for about 16 percent of a typical hotel's water use, second only to landscaping. Every 10 kilograms of laundry uses about 50 gallons of water.[7] There's a lot of water we can save. Now, scratch beneath the surface, and you'll start to question the why behind these campaigns. By consumers saving water, the hotel is saving money. The American Hotel and Lodging Association notes these programs can save a hotel up to 17 percent on associated costs.[8]

That's a massive benefit to their bottom line. Think about the other ways these hotels could save the Earth if they genuinely cared. How about installing water-saving toilets or showerheads? How about sustainable landscaping or farm-to-table food products? How about not producing millions of little plastic bottles filled with soap and shampoo—usually parked right next to that plastic sign telling you to reuse your towels, mind you—opting instead for a single refillable container in the shower? Or, how about stopping wasting paper and plastic printing up the little placards telling us they care about the Earth? There is so much that hotels can do. Instead, they expect you, the customer, to do the heavy lifting.

There's a laundry list (pun intended) of companies being caught with their pants down. Chevron's famous 'People Do' environmental campaign of the 1980s came at the same time they were violating the US Clean Air and Clean Water Acts. In 2017, Walmart had to pay US$1 million to settle claims its environmentally friendly product descriptions misled consumers. Most notably, the 2015 Emissionsgate scandal embroiled German automaker Volkswagen. The company tried to refute adverse claims against diesel fuel, announcing the launch of new technology to reduce vehicle emissions. Analysts found VW intentionally programmed these emission controls to only work during government inspections. Federal agencies made Volkswagen pay US$14.7 billion in restitution, indicting ex-CEO Martin Winterkorn on fraud and conspiracy in the United States.[9] Imagine what could have happened if they applied that technological know-how and innovative thinking to a good cause!

To try and combat greenwashing, while increasing transparency, stakeholders began to ask for corporate sustainability reporting. Early on, these reports worked to get under the hood of sustainability efforts. Eventually, though, the system became bloated, broken, and more of a ticking-boxes exercise. Instead of providing the hard data readers needed to make an informed decision, reports became glossy magazines filled with pictures of smiling children. There's now more space spent talking about charity (usually in the form of forced employee volunteerism), messages from executives (usually templated from year to year), and corporate credentials (usually to sell more of their products).

Those reports that do provide information tend to either incite fear, given how long they are, or inundate readers with numbers. Water effluent performance, energy efficiency, worker empowerment programming, office recycling data, employee volunteerism, supplier audits, revenue, profit, emissions, numbers, numbers, numbers. Spreadsheets in landscape format, with tiny numbers bleeding together into a black blob on the page. Lists upon lists of performance indicators. In trying to be more transparent, these reports are doing the opposite. These tell readers a whole lot of nothing. Complicating matters are multiple global standards that make for messy reporting. In a recent McKinsey poll, nearly 70 percent of investors said there should only be one standard to avoid 'inconsistency, incomparability, or lack of alignment'.[10]

All that brings us straight back to the supermarket shelf. Now, greenwashing is easier than ever. Why? First, more products and messaging bombard us every day. Today's consumers can choose between 40,000 to 50,000 unique products when they go to the store. That's up from just 7000 in the early 1990s.[11] Store size has also doubled over the past decade, giving brands more space to push their stuff. That means brands have to do more to stand out from the crowd and get your attention.

That leads us to the second way greenwashing is becoming easier: technological advancements. A dirty little secret of the FMCG industry (that's 'fast-moving consumer goods' for the uninitiated) is that product development, package design, and even placement on the shelf rely on sophisticated technology. With the advent of social media, marketing teams now have access to more of your information than ever before. You know when you ignore the cookie alert on a webpage? It's those cookies that help companies track what you're doing. All of these little bits of data become what ultimately ends up on the supermarket shelf. There's now even predictive AI that can tell you what you like before you even know it. Billions are spent every year on this because *you* are the product at the end of the day.

And what is it you care about? That's right, the Earth. Further exacerbating the issue of greenwashing is the exponential growth of companies claiming to care. Of course, some may have altruistic intentions. Most, though, are simply capitalising on the green wave to keep their necks off the chopping block. Nowadays, a private sector enterprise has little social capital if it doesn't talk about how sustainable it is. We've gone beyond the point where companies consider sustainability a 'nice to have'. Now, it's imperative.

Taken together, this puts us smack bang in the middle of a perfect storm. The inundation of new products, technological sophistication, and our own desire to be green are perversely working against us. It's like we're gleefully listening to the band on the *Titanic*, ignorant the ship is sinking right beneath our feet.

Consider this book your lifeboat.

Chapter 2

How to Spot Greenwashing

Luckily, corporations aren't as smart as they think they are. As insidious as it is, greenwashing can often be very easy to spot.

Modern-day greenwashing comes in many forms. Firstly, you've got your blatantly false claims, like how Chevron continues to push how environmentally friendly it is. Then, there are misleading labels that do little to support their claims. Some companies swap out one good for one bad, like sweeping the child labour in Bangladesh under the rug because you've given to a children's charity in Zimbabwe. Irrelevant claims also pop up, whereby company X will say their products don't have a specific chemical but fail to mention that regulations have already banned that particular chemical. They didn't do anything themselves. They're simply following the law. Last, but not least, are products doused in the colour green. Just because it looks environmentally friendly on the package doesn't mean it is. How dumb do they think we are?

The US Federal Trade Commission has a series of guidelines on deceptive green marketing claims to educate consumers on greenwashing. These guidelines cover everything from environmental benefit claims to offsets, copy stating something is free-of or non-toxic, and what it means if something is renewable. They also lay out a few interesting examples of greenwashing to get consumers thinking critically:

> A plastic package containing a new shower curtain is labelled [sic] 'recyclable.' It is not clear whether the package or the shower curtain is recyclable. In either case, the label is deceptive if any part of the package or its contents, other than minor components, cannot be recycled.

> A trash bag is labeled [sic] 'recyclable.' Trash bags are not ordinarily separated from other trash at the landfill or incinerator, so they are highly unlikely to be used again for any purpose. The claim is deceptive since it asserts an environmental benefit where no meaningful benefit exists.[1]

You see, greenwashing isn't always so easy to spot. It takes much more than just looking at marketing copy and calling BS. There's an entire ecosystem to take into account. Yet when you Google 'greenwashing', perhaps in an attempt to educate yourself, things can get even worse.
Google 'greenwashing'.
Go ahead. I'll wait.

What'd you find? I'm guessing a very confusing assortment of examples, activists and journalists decrying how bad it's gotten, as well as some version of the 'seven deadly sins' of greenwashing. There's no shortage of marketing and economics experts with their own spin on greenwashing sins to avoid. Even more are continuing to add to the pile. Don't worry—I'm not trying to be one of them. I want to compile all of these definitions into something much more manageable. When you dig into all of this, you have a lot of overlap, with people saying the same thing in slightly different ways.
Ultimately, I landed on three big buckets of things consumers

can look for to identify greenwashing. The first is anything involving 'green speak'. That includes the fluffy language and misleading labelling littering packaging in your local grocery store. The second is misdirection. Greenwashers are excellent at distracting consumers from real issues. Consumers, for their part, are just like cats with shiny objects. Don't be a cat. Lastly, an entire category is related to false hope, broken promises, and fearmongering through greenscamming.

Oh, and by the way, I absolutely refuse to give these a cutesy name like the *holy trinity of greenwashing*. It might look good on a brochure or as dot points on a presentation slide. In my opinion, these monikers trivialise such an important topic.

Let's look at each of these three buckets in turn.

Green Speak

During an average week, Melbourne's busiest train station, Southern Cross, receives 1.2 million passengers into and out of the city. The train network brings people in from across the metropolis, while regional trains spit out those from further afield. With so many stepping foot through the halls, not to mention those waiting for a departing train, it's a paradise for marketers.

I step onto the escalator that leads down to street level, heading to an appointment in the CBD. The open-air entry hall has a vaulted ceiling a good two stories high. Directly in front of anyone on that escalator is space for a massive banner. We're talking nearly an entire story tall and just as wide. It's probably the most significant single piece of marketing real estate in the station, one I'm sure PR teams covet.

Usually, I ignore whatever banner might be taking up the space. On that particular day, though, it caught my attention. It was an advertisement for a major Australian bottled water brand. On one side was a metal flask, similar to the kind eco-warriors carry for water. The copy above read, 'when you can'. Juxtaposed on the other side was the image of their new eco-friendly bottle with the simple phrase, 'when you can't'. The apparent insinuation was that their new single-use plastic bottle was just as environmentally friendly as a reusable thermos. It's a great piece of marketing, but was it just plain greenwashing?

Most of us rightly associate greenwashing with communications or marketing. We look at advertisements, like this brand's campaign, and take at face value messaging around reducing waste. We read packaging copy, inundated with so much jargon it's impossible to decipher. We also take grandiose environmental claims for granted. There are laws against false advertising, so what we're reading must be true.

Doing a bit of digging, the water brand's campaign actually seemed to check out. The brand in question, Mount Franklin, sits under the Coca-Cola Amatil Group. As part of Coke's global efforts to create a 'World Without Waste', the majority of Mount Franklin bottles are now made from 100 percent recycled material.[2] Label messaging boasts about this worthwhile achievement while also encouraging Australians to recycle the bottle again when done. Clever marketing, a great initiative, and consumer engagement toward eco-friendly practices make for a solid campaign.

Unfortunately, Mount Franklin's campaign is likely the exception rather than the rule. For many companies, engaging in green speak is just another tool in their arsenal to sell, sell, sell.

Green speak occurs in several ways. The most obvious are outright lies told to consumers. This book is full of examples, so I won't waste space on any just yet. Greenwashing is itself an exercise in fabrication, with these lies being just the tip of the iceberg. Without the proper regulatory controls, marketers have a lot of leeway to interpret what's appropriate. As consumers have caught on to these lies, this approach is being tossed aside for more clever deceit.

What kind of deceit, you ask? Often, companies will make claims but offer no proof. Marketers are betting you won't be bothered fact-checking. How many products do you see daily claiming they're eco-friendly, green, or sustainable? How many have you actually looked into to find out the truth? As we've come to find out, the marketers are right.

When brands do offer proof points, these can be so vague they lose meaning. Statistics are a great example of this. Claiming to be 50 percent less polluting, 20 percent more sustainable, or 99 percent recycled is hard to interpret without a relevant comparison. Another greenwashing example from the US Federal Trade Commission demonstrates what we mean by these claims.

> An area rug is labeled [sic] '50% more recycled content than before.' The manufacturer increased the recycled content from 2% to 3%. Although technically true, the message conveys the false impression that the rug contains a significant amount of recycled fiber.[sic][3]

If claims aren't vague, they can be full of jargon, gobbledygook, or fluffy language. Brands use this method to demonstrate an air of superiority while inciting confusion among readers. The copywriter just has to show how their product is different from and better than the competition. They chuck in a bunch of terms that sound good but are difficult to understand. You'll assume they are more advanced, safer, or better. Clearly, they seem to know more than you, so why question their authority?

Chinese advertising takes this to the next level. While living in Shanghai, one of my friends told me about a side gig. The role was, and I'm not making this up, to play a doctor in a television advertisement. Was he a doctor? Absolutely not. Was he a middle-aged white guy? Tick and tick. When he arrived on set, they gave him a white lab coat and told him to stare intellectually at a beaker. A couple of hours later, and a few hundred RMB richer, he left with a great story to tell.

I never did see the final advert, but there were plenty like it floating around. You can probably guess what was going on. Someone portraying a doctor in the advertisement lent the product an automatic gravitas. No viewer would go to the trouble of looking up my friend, meaning they'd never find out he wasn't a real doctor. Although not as in your face, greenwashing uses some of the same tricks.

Lastly, some companies proudly boast they adhere to specific business or operational practices. They insinuate this makes them best in class. They might fail to mention that these practices are legally mandated. Imagine a fast food restaurant going on about how good they are for paying employees eight dollars an hour when that's the minimum wage. Or a nuclear facility boasting about their procedures to prevent meltdowns. They have to have this in place. It's not a point of differentiation!

Chevron used this tactic in their 'People Do' campaign mentioned above. The company crowed about their list of good deeds, like flora and fauna preservation programs. Eventually, people pieced together these good deeds were, in fact, regulatory requirements. It wasn't that Chevron wanted to execute these programs. It's that they had to if they wanted to stay in business.

That's probably a lot to digest. At the end of the day, how can you ensure you don't fall victim to green speak? Ask yourself if what you're reading on a pack, or seeing in an advertisement, could stand up to scrutiny. Honest companies will tell their story with facts and examples in a way anyone can understand. Desperate to be seen as green, dishonest companies will try too hard. It's like the corporate version of love bombing, and it's just as abusive.

Misdirection

After all the trials and tribulations along the Yellow Brick Road, the only thing Dorothy and her coterie wanted to do was see the Wizard. They battled the Wicked Witch and made it into the Emerald City. Now the foursome stood before the Great and Powerful Oz. Giant pillars of fire light up the room, framing a great translucent head barely visible through all the smoke. He bellows out commands and demands subservience. The group is absolutely petrified but make their requests nonetheless.

We all know what happens next. While the group is shaking in their boots, little Toto scurries over to a green curtain in plain sight the whole time. He grabs it in his mouth and pulls it open, revealing the Wizard's true identity. A feeble old man, opposite in every way to the literal figurehead he manifested, is busy pulling levers and turning knobs. Noticing the jig up, he famously exclaims, 'Pay no attention to that man behind the curtain.' By then, though, the ruse was over.

This theatrical example of misdirection holds so many parallels to the way corporations engage in greenwashing. They portray what they're doing as grand and altruistic when these are just feeble attempts at keeping power.

Even after the discovery, Oscar Zoroaster Phadrig Isaac Norman Henkle Emmannuel Ambroise Diggs (yes, that's his actual name in the book) tries to keep his scam going. How often do companies try to keep their subterfuge alive after the public has caught on? We have oil companies positioning themselves as environmental activists, fast fashion retailers pushing small-scale recycling programs, and airlines graciously letting passengers pay to offset their emissions. All of this is misdirection in its truest sense: look over here, not over there.

A classic example of misdirection comes to us from Westinghouse Electric. In the 1960s, the company released a rather dubious ad campaign just as environmentalism was taking off in the United States. They showed a pristine lake with one of their giant nuclear power plants in the background. The strapline read, '[w]e're building nuclear power plants to give you more electricity.'[4] They go on to espouse plant safety. Never mind that a reactor meltdown in Michigan had almost wiped Detroit off the map only three years prior.

Companies today employ many of the same time-honoured tactics to misdirect the public. The first of these comes in the form of selective disclosure. Like Oz didn't mention he was a snake oil salesman from the Midwest, many companies conveniently keep what they talk about tightly bookended. Some may rightly market a single product that is truly environmentally friendly. Yet, they don't tell you the rest of their products aren't nearly as green. Others brag about how they're best in class versus the competition. Never mind the fact their entire industry is highly polluting. A third approach is to only talk about the positive aspects of a product or service. All the nasty stuff is quietly swept under the rug.

Another popular tactic in misdirection is what I like to call *having imaginary friends*. You probably would have seen some of these imaginary friends stamped onto the front of packages, especially in the food aisle. These are eco-labels certifying a product as sustainable, eco-friendly, or ethically sourced. Some of the most popular are from the Forest Stewardship Council, UTZ, or the Rainforest Alliance (you know, the one with the cute little frog). The Ecolabel Index tracks these different green stickers, with a current repository of 455 eco-labels in 199 countries across twenty-five industries.[5] That's a lot of labels, each with their own standards and methodologies.

With so many floating around, it's getting harder to police and ensure credibility. In some respects, the foxes are guarding the henhouse. That's because many of these labels are not from reputable third parties. Industry associations, or sometimes even a single company, created them to add a green sheen. You read that right. Some companies make their own eco-labels to stamp on their products. Talk about patting yourself on the back.

We look at these labels and think a company must have its house in order. Otherwise, how would they get this seal of approval? Unfortunately, like selective disclosure, eco-labels do not guarantee much of anything. A 2021 Greenpeace report[6] looking into forestry and palm oil certifications notes eco-labels do not stop deforestation, human rights abuses, or indigenous land grabbing. They also do not necessarily improve the livelihood of smallholder farmers or biodiversity in tropical regions.

The report sadly quotes Unilever's Chief Supply Chain Officer saying, 'Certification does not equal the definition of deforestation free.'

Researchers examined nine certification schemes' environmental, human rights, and labour records. Some, like UTZ, were prevalent global names. Others were industry associations, including the Roundtable on Sustainable Palm Oil (RSPO). The findings are damning. The Environmental Investigation Agency found, 'Non-adherence to the RSPO's standards is systemic and widespread, and has led to ongoing land conflicts, labour abuses and destruction of forests.'[7] Other investigations found no significant difference in sustainability performance between those plantations with eco-labels versus those without.

Ultimately, the Greenpeace report concluded:

> ... certification enables destructive businesses to continue operating as usual. By improving the image of forest and ecosystem risk commodities and so stimulating demand, certification risks actually increasing the harm caused by the expansion of commodity production. Certification schemes thus end up greenwashing products linked to deforestation, ecosystem destruction and rights abuses.[8]

The final approach to misdirection is a bizarre one. In fact, I'm shocked companies and PR agencies still think people are stupid enough to fall for it. This misdirection involves employing targeted semiotics, which is the use of symbolism to portray meaning. Essentially, they're trying to trick you with pictures. The Westinghouse advert is just one example. Companies today love to use forest, ocean, or mountain backdrops to show they're sustainable. They also bluewash, overstating their commitment to social issues. Pepsi did this in its infamous advert showing Kendall Jenner solving racism and police brutality with a soda. This ad lasted less than twenty-four hours before being pulled due to public backlash.

For the entire decade I lived in Shanghai, the city was constantly under construction. The clanking of metal on metal, guttural pangs of jackhammers blasting concrete, and the rumble of trucks at all hours became white noise. One thing I found amusing was the walls put up to cordon off construction sites. Some were white and boring. Others, though, were works of art.

Across from my old apartment was a multi-square-block hole, home to a future high-rise tower and mall. You'd have to wear a face mask passing by to avoid inhaling the massive amounts of toxic dust kicked up from machines. Construction is considered one of the world's most polluting industries. Looking at the twelve-foot tall walls in front of the site, you wouldn't know that. They were covered in campaign posters telling passers-by to reduce their carbon emissions. The iconography of whales, birds, and even pandas was juxtaposed against a thermometer bursting from rising global temperatures. Whoever designed and approved this clearly had no sense of irony.

The difficulty in identifying misdirection is that one must do quite a bit of research to pinpoint the truth. By its very nature, selective disclosure is just that: selective. Take this example. A 2018 statement by Starbucks noted they would '... stop using disposable plastic straws by 2020, eliminating more than one billion straws a year.'[9] I first encountered these new straw-less lids at my local Xujiahui Starbucks in Shanghai and immediately snapped a picture to post on my social media networks. What a step in the right direction for a company growing by thousands of stores a year in China. There really is no need for straws in your iced coffee, so I saw eliminating them as a great move.

If you've yet to use one of these lids, it's essentially a grown-up version of a toddler's sippy cup. I'm constantly worried I'll spill or splash coffee all over my shirt since the hole cover tends to pop out of its socket. However, this was a small price to pay for a more sustainable world. As I started to delve deeper into the science behind the lid, I found a lot of grumbling from my professional peers. Their studies have shown the new tops use more plastic than the original lid and straw combination. While most were busy congratulating Starbucks on a job well done, the company was contributing to the problem they touted solving. That's selective disclosure.

So, what's one to do? I'll not lie to you and say it'll be easy. Some things, like suggestive or symbolic imagery, are not too hard to identify. You can also probably spot when a company from a polluting sector tries to say they're the best of a bad bunch. The lesser of two evils is still evil.

When it comes to selective disclosure and stamps of approval, though, don't take anything at face value. I'm not a cynic, but I suggest you not trust anything until you've vetted it yourself. That means using all the tools at your disposal, including the internet, corporate reports, and reputable NGO statements. Common sense plays a big part, too. The next time Shell or BP try to sell you the positives of their environmental record, your gut and brain should be sounding warning bells. Like Toto, don't be afraid to pull back the curtain.

Greenscamming

If you thought companies lying on their packaging or misdirecting you with a few dodgy stats was terrible, just wait until you hear this. Some cashed-up companies, organisations, and individuals are known to go out of their way to actively discredit sustainability. These actions go far beyond simple greenwashing and enter the territory of greenscamming. Greenscamming is a concerted effort to capitalise on sustainability by bringing about its downfall. Any information, science, or people getting in the way of achieving a particular end goal are hostile combatants. As you'll soon see, the lengths greenscammers will go to boggles the mind.

The poster child of greenscamming is the fossil fuel industry. For years, they have called into question the validity of climate change and climate science. Only recently have we discovered just how deep this conspiracy goes. In the book's next section, we'll dive deep into the insidious work of ExxonMobil. For decades, the oil giant invested squillions into denying links between fossil fuels and climate change. This, even though executives were well aware of the impact their business had on the environment back in the 1980s. ExxonMobil had gone so far as to launch a media campaign aimed at high school science teachers, emphasising 'uncertainties' in climate science. Fossil fuel companies also continue to fund anti-climate lobbying groups. One, in particular, has run a decades-long smear campaign against a single lawyer who dared take them on.

Peeling back the layers even further, we find examples of astroturfing popping up worldwide. Astroturfing is a practice that adds credibility to front organizations, including those supposedly dealing with sustainability, environmentalism, and climate change. If your group is trying to drill for oil in a protected region, you can't go to the local council and say as much. The public won't be too happy if a group called 'Drilling for Dollars' starts setting up shop. Instead, astroturfing organisations will create more eco-friendly names hiding their true intentions.

They also have very opaque funding sources. For example, ExxonMobil has been behind funding over one hunderd climate-denying astroturfing organizations. Notable families, like the Waltons and Koch Brothers, are also known to invest millions into scam groups. A 2013 investigation[10] found the Walton's gave $6.3 million to a Los Angeles group called Parent Revolution. While its '… stated goal is to empower poor parents to transform their low-performing neighbourhood schools into successful ones' in reality, the group is a front for '… private charter school operators and their wealthy allies in the [ultra-conservative] Education Reform movement who support weakening and ultimately abolishing the teachers' unions.' They want to replace public education with for-profit schooling.

Unless one does their research, there is really no indication on the surface of how nefarious these groups are. Look at some of these names compared to what they actually do.

- The European Institute for Climate and Energy is a media outlet publishing fake climate news.
- The non-profit Center for the Study of Carbon Dioxide and Global Change are staunch believers global warming will benefit humanity.
- Since 1989, the National Wetlands Coalition has worked to thwart US wetland protection policies and water down the Endangered Species Act.

National players sometimes get involved as well. Russia is known to drum up anti-fracking sentiment to better position its European oil sales. Several groups close to conservative Canadian politicians also push a message of 'ethical' oil that will supposedly benefit Canadian society. The infamous *wu mao*, an army of nearly 300,000 trained Chinese netizens, engages in posting state-sponsored pro-China social media commentary. These groups are capitalising on and exacerbating a world where information integrity is quickly becoming an outdated concept.

For most of these groups, the goal isn't necessarily to fully realise any set ambitions. China's netizen army is no more likely to stamp out all negative online sentiment than I am to get six-pack abs. Ethical oil politicians in Canada know they're fighting an uphill battle. Any concerned citizen with a bit of research know-how can even thwart those groups with false monikers.

Instead, all of these groups aim to raise doubt. In George Monbiot's book, *Heat*, he quotes a leaked memo from a tobacco company. In it, they explain well the motivation of all greenscamming organisations: 'Doubt is our product since it is the best means of competing with the "body of fact" that exists in the mind of the general public. It is also the means of establishing a controversy.'[11] One doesn't have to look too far to find establishing controversy is a lucrative business these days.

But you're smarter than that. I know you're not hopping on Twitter and believing the latest conspiracy theory. And since we're talking about saving the planet, I'm assuming you're not a flat-earther either (fingers crossed). Where once it might have been easy for these groups to pull the wool over our eyes, we now have access to more information than royals only a few hundred years ago. The key is that we have to use it. Technology is your friend, especially in the fight against greenwashing.

Chapter 3

More Than Just Green

Over the past forty years, greenwashing has evolved in both understanding and practice. Today, most define the term greenwashing as deceptive marketing applicable to the environmental impact of a product, brand, or company. It's basically showcasing the eco-friendliness of your brand. New terminology, like 'bluewashing', relates to a company's social impact. In this context, bluewashing shows how ethical your company may be. Do you have human rights abuses or forced labour in your supply chain? How does the production of a pair of your jeans affect waterways and the ocean? What impact do your products have on local communities?

That last question is something most companies forget. Production doesn't operate in a vacuum. There are impacts beyond the office parking lot or factory walls. Local communities and stakeholders are just as much part of a company's sustainability story as a recycling program or emission data.

A story that brings this to light comes from lush, semi-tropical Guangxi. Located in China's south, the panorama spills over with limestone cliffs, crystal turquoise water, and canopies of full, pencil-thin trees. It's the perfect spot to relish China's pastoral past. There's no better place to set up shop for a paper mill.

That's precisely what Stora Enso, the world's oldest company, decided to do. Better known locally as Asia Pulp and Paper, Stora Enso is one of the largest paper manufacturers. Today it produces paper, packaging, wood products, and biomaterials. In 2010, the company set its sights on Guangxi as the location for its newest mill. On paper, Stora seemed to tick every box. They consulted with provincial and local government officials, strategised how to build their operations sustainably, and met all building regulations. As they came to find out, though, they forgot one key stakeholder.

Forestry is not a new industry in Guangxi. The local communities have engaged in tree felling and production for hundreds of years, with the manufacturing base and supply chain established well before Stora Enso arrived. Livelihoods for generations of workers depended on this most central of provincial industries. When Stora came in, they threatened to upset the natural ecosystem through large-scale farming and putting many families out of work. It was only as the mill walls began to rise that locals realised what was happening.

What happened next seemed ripped from the scenes of a horror movie. Local workers, armed with machetes, began taking matters into their own hands. Instead of chopping down trees, they were going after representatives from Stora Enso. Government officials viewed as capitulating to profit and greed were also targeted. Even when trying to rectify the problem through town hall meetings, tensions boiled over into fistfights. Over a decade later, Stora Enso's reputation in the region is less than stellar. While the mill is operational, the journey taken to get there is the stuff of nightmares.

While horrific, one can consider the Stora Enso incident a parable on motive. The underlying reason for building the mill wasn't to run from machetes. Instead, a culmination of little oversights resulted in a massive problem. The company has also learned from these mistakes. They've spent the past decade engaging local and national stakeholders as they continue to repair the breach of trust.

Why a detour into the forests of China? Motive is important to consider as you stare at shelves of products all claiming to be green. As we'll see throughout this book, plenty of bad actors exist. When they develop messaging, they know precisely what they're doing. These folks know greenwashing as well as any expert. Except they are conscious of how to play the game, so consumers don't notice.

But there's also a second category of greenwashers. These are companies that have mistakenly done something untoward. With more and more companies realising both the merits and financial benefits of sustainability, there are going to be heaps starting their eco-journey. One of their marketing managers may have heard a few good catchphrases from a colleague at another company. They introduce it to a senior executive who green lights adding the fluffy language to their messaging. At no point was anything vetted. Someone made assumptions and jumped on the bandwagon.

Or maybe the people in charge don't know enough about sustainability to make informed decisions. Without a consultant or expert's guiding hand, there are likely to be missteps. They may not realise that while their environmental record is clean, issues across the supply chain call the organisation's legitimacy into question. Perhaps they received a stamp of approval from a dodgy certification bureau. As is often the case with sustainability, said company may have handed responsibility to some junior staff members. Mistakes are bound to appear without oversight, stakeholder buy-in, or proper review mechanisms.

With this second group, I can be a bit more lenient. They aren't knowingly trying to pull the wool over anyone's eyes or misdirect from something more sinister. Of course, the negative impacts of greenwashing are ultimately the same regardless of motivation. But at least this second type of company had its heart in the right place.

At this point, you're probably wondering why in the world you should care about all this. There are a few big reasons that go well beyond just being green. The biggest of these is because they're lying to you. Who likes that? Not me! What's worse is these companies are charging you for the pleasure. It's like the world's biggest Ponzi scheme where you're the lackey. By positioning themselves and their products as sustainable, these companies can also typically command a price premium on what they're selling. So now you're not only paying for them to lie to you—they're charging you even more in the process.

Beyond just how grating being lied to is, greenwashing has other impacts, too.

When it comes to consumer sentiment and brand loyalty, people often gravitate to products perceived as 'good' for the user or the Earth. As we've discussed, companies know this. It's the foundational rationale for greenwashing. Of course, there's a balancing act happening here. Messaging should allow for greater brand loyalty but not go so far as to be seen as the lie it is. Companies caught in greenwashing lies often have a tarnished reputation, even if only temporarily.

More and more, people are looking for a place of work that has a purpose beyond just selling stuff. Sustainability professionals have discovered that employee engagement, retention, and productivity go up in purpose-driven organisations. Corporate psychologists are beginning to identify the negative impact greenwashing has on this esprit de corps. Unsurprisingly, a 2022 Portuguese study found that '… employees' perception of greenwashing relates negatively to their career satisfaction, organizational pride, and affective commitment.'[1]

To put this another way, employees are embarrassed when they find out their company is engaging in greenwashing. For anyone who's ever hated their job, you know just what kind of impact this has on your productivity. While the list could go on and on, there is one other negative impact of greenwashing I just have to point out. Whether you realise it or not, falling victim to greenwashing makes you an accomplice to all the terrible things these corporations are doing. Yep, that's right. You're a culpable enabler funding oil slicks in the ocean, coltan mines in the Congo, and even the melting sheet of ice that poor polar bear is floating on. As we'll see, greenwashing (or rather, green gaslighting) is another way for climate denialists to get their way. While people are busy focusing on the 'positive' things they say they are doing, corporations are busy plundering, polluting, and profiteering just out of sight.

Okay, John. So I totally care about this, but it's not like I'll be able to do anything about it. I don't work as a marketing exec at a Fortune 100. My daddy or mommy isn't the CEO of a big conglomerate. I'm just a lowly consumer having these products forced upon me without any recourse or choice. Right?

Wrong.

You're consciously deciding to change the paradigm by simply picking up this book. Reading between the lines and identifying when companies, politicians, and individuals are lying to you will go a long way towards ending greenwashing. Will it solve everything? Of course not. This understanding must be coupled with action, particularly as consumers begin to speak with their wallets and voters at the ballot box. Companies like money, and politicians like power. The second we begin to chip away at this, you can guarantee we'll see change starting.

Over the following few hundred pages, you'll become equipped with the tools needed to be the proverbial change. We've already talked about the various types of greenwashing you can encounter. While this might make sense on paper, seeing it in action is really telling. Much of this book will look at historical and contemporary examples of the what, how, and why of greenwashing. I've very consciously tried to avoid this being just a book of case studies. If you want that, enrol in a boring business school course. Instead, I'm trying to add to the already broad research involving greenwashing by arming you with actionable things to counter all the negative you might come across.

The book will flow over three large sections covering **corporate**, **public sector**, and **individual** greenwashing.

We'll start by looking at the **corporate sector**. They are the biggest group and most likely to push overt greenwashing into our everyday lives. But the corporate sector is multifaceted. That's why I'll break this section down into three big segments. The first deals with those I like to call the 'unsustainable'. These are companies that cannot be sustainable no matter how hard they try. Defence, tobacco, and oil all fall within this category. While they certainly have a negative environmental impact, what's especially interesting is their record on human rights. When considering the unsustainable through this lens, all the greenwashing in the world just doesn't hold up.

The second group is those companies that are bad for the planet but still have the wiggle room for positive change. These aren't lost causes like those in the previous section. Airlines, plastics, and chemical companies may prove a scourge to the Earth today. But a little bit of consumer pressure, and some novel innovation, may prove just right to force a more environmentally friendly evolution.

Given its size, scope, and impact, I've separated the fast fashion industry into a third section. In it, I cover the industry's evolution from antiquity to today. I also look at some of the most troubling incidents marring this history. Along with calling out the worst of the worst, this section also explores some of the positive changes the vanguard is making to position the fast fashion sector for a more sustainable future.

Next, we'll explore how **public sector** organisations, international institutions, and even individual nations all practice their own unique form of greenwashing. Although you might assume groups like the United Nations are committed to positive impact, the results of conference after conference tell a different story. This is even more demonstrative when we examine how greenwashing can be a political goldmine.

Our final big section will consider some of the major **individual influences** in our lives. These influences can set us on a positive trajectory or put us into a viciously negative cycle. Unfortunately, we haven't been going in the right direction when it comes to sustainability and greenwashing. I highlight three segments of influence that can make or break our efforts to build a better future. Activists, moralists, and non-governmental organisations, while they might mean well, have become hamstrung by their lack of pragmatism. Then we get into the individual influencers like the ultra-wealthy, celebrities, and climate narcissists. Finally, we put a mirror up to ourselves and our actions to see what we can improve.

Before we dive in, I want to point out one group curiously missing from these pages. These are the financiers of greenwashing and bad corporate behaviour. Sure, corporations are profiting off all of this. But even the most prominent organisations need investment. Take, for example, the banks that finance the fossil fuel industry. JP Morgan Chase and Citi Bank are just two of the biggest investors. How about those with nuclear weapons and defence as part of their portfolios, like Australia's largest pension fund, AustralianSuper? Panning out, we could point to how a select group, including Blackrock, essentially funds the majority of corporations around the world.

While their portfolios might have highly destructive sectors, these investors may also fund several sustainable projects, too. What would happen if AustralianSuper just shut down overnight? Sure, the nuclear industry would lose a big chunk of change, but so would millions of older Australians.

Does this make these financiers culpable? Perhaps. Does it make them ultimately responsible for what their investments do? I'm not so sure. While some of my more activist colleagues have called on banks to divest from investments in polluting or unsustainable industries, this is easier said than done. One concept we'll look at later is that of a 'just transition'. For change to happen, there must be a fallback for employees, suppliers, and other key stakeholders. Capital investment in infrastructure is also critical. What happens to workers if banks stop providing funding and a company collapses? Ensuring a positive outcome for both the planet and its people makes this a tricky situation.

Adding to the complexity is the secrecy of most funding sources. While a company may be public, that doesn't mean we necessarily know where they're getting their money. Piecing together the tangled web of shell corporations, tax havens, and other forms of monetary manipulation is not something I have the resources (or interest) to do. Plus, it would probably put you to sleep. Given these murky waters, I've decided to leave the money conversation out entirely.

Throughout this book, I've tried to address some of the most foreseeable questions, include the most relevant insights, and provide the most practical actions that you, dear reader, can take. While I'm sure I won't cover all comments, questions, and concerns, I bet most of you fall within one of three segments. You might be coming to this topic with just a bit of greenwashing knowledge but a lot of passion for doing the right thing. Others, like myself, might work in sustainability each and every day. A third group may not realise it yet, but they are the central actors I talk about in these pages.

Each of you will have different expectations from this book I hope I can meet. As you move along this journey, think of how the various bits of information, anecdotes, and cases apply to your personal or work life. Consider the many daily examples you might encounter that haven't made it into this book. Most importantly, consider how you can do better since protecting the planet is a team sport.

Ready? Set? Go!

PART 2

THE CORPORATE SECTOR

Chapter 4

The Good, the Bad, and the Ugly

Corporations have a pretty shit reputation when it comes to sustainability. Their track record is, well, less than stellar. Watch. I'll show you.

I'm obviously sitting at my computer right now, so I'll open up an internet browser and jump over to Google. In the search bar, I'm typing the pretty broad string: 'corporations AND negative impact'. Bam, 2.4 billion results. The first page alone has such damning titles as 'Multinational Corporations Do More Harm Than Good' and 'Multinational Corporations: Good or Bad?' Let's go one step deeper and search for, say, 'corporations AND negative sustainability impact'. A more manageable 220 million results. 'Corporations AND negative environmental impact', 206 million results. 'Corporations AND negative social impact', 1.9 billion. Many of these pages mix the bad veiled within the good. For example, the page title '20 Companies With the Most Impact' buries the poor performers under a list of those with a positive impact.

Let me take off my sustainability nerd hat for a second and search like an average person: 'Companies that have a negative impact on the environment'. A jaw-dropping 5.6 billion results. What about 'Companies that have a negative impact on the planet'? Even higher at 7 billion. As a bit of context, a Google search for 'god' only turned up 5.4 billion results; 'Elon Musk' 3.8 billion; and 'Donald Trump' 2.7 billion. There is more information on corporate sustainability performance than the guy who bought Twitter.

Right at the top is a blurb from *Big Issue* naming Coca-Cola, Pepsi, and Nestle as companies with the world's worst waste records. Scanning the list of titles, I also spot a list of the worst companies on the planet, energy companies destroying the Earth, and a list of firms concealing their negative environmental impact. Not exactly the mixed bag we saw before, but indeed more representative of the reality.

Clearly people are concerned about the state of our planet and the impact corporations are having on it. Can you imagine if the internet existed in the early 1900s? Jacob Riis, a social reformer and muckraker working in the United States in the early twentieth century, would probably be publishing blog after blog after blog. Much of his work focused on the plight of workers in the impoverished Lower East Side of Manhattan. He also used his photographic skills to document child labour, unsavoury workplace conditions, and government corruption. His seminal work, *How the Other Half Lives: Studies Among the Tenements of New York*, brought to light many of the social issues of the day and began a period of national reform.

Those immigrants, if they could make it past inspection on Ellis Island, would probably end up in crowded, rat-infested tenements on the Lower East Side. They were likely to spend their days in a hot, cramped, and dangerous warehouse, operating machines that could quickly turn into torture devices for misplaced limbs. If they did injure themselves, they could attempt to rely on the charity of others. Yet, given how stretched most budgets were during this era, it was probably the end of their American dream.

Sure, we're not still living in a Jacob Riis or Charles Dickens' novel. At least not in the western world. That's because a lot of the nasty stuff we would have seen in Chicago or New York's Lower East Side is happening in the developing world. Nothing demonstrates this better than the management of recycling. Fellow Shanghai ex-pat, Adam Minter, has done a great job at summarising just how much trash ends up this side of the Pacific. His book, *Junkyard Planet*, is part exposé, part environmental call to action. Published in 2013, Minter travels across Asia, following the trash trade to see where it goes. He starts with the humble Coca-Cola can, but eventually finds his way into everything from electronics, to cars, to all manner of scrap metal.

Most of the stuff you throw away, recycling or not, is probably loaded up onto a cargo ship and sent across to Chinese cities you've never even heard of. Experts estimate that in '… the U.S. alone, nearly 4,000 shipping containers full of plastic recyclables a day had been shipped to Chinese recycling plants.'[1] A full 70 percent of the world's plastic waste ended up somewhere in China as of 2017. We're talking millions of tons a year. The Chinese were more than happy to buy these recyclables as they could cheaply take them apart and reuse the materials.

This recycling approach has adversely impacted both the environment and the people living within trash-sorting hot spots. The stuff you so diligently (or, maybe, not so diligently) separate in your suburban model home will eventually get separated further by the hands of a scrap sorter somewhere in Asia. Minter recalls his first trip to Foshan, one of China's centres for the scrap trade.

> Meanwhile, over in the farthest corner of the yard, the flicker of flames might send black smoke into the not-quite-as-dark night. The smell would be noxious (and, depending on the wire, dioxin-laced), but the goal would be anything but: profit. Wires too small to run through the stripping machines were a favourite item to burn, but anything would do if copper demand was strong; in the morning, the copper could be

swept out of the ashes. One night, I recall clearly, I saw a row of a half dozen electrical transformers—the big cylinders that hang on power lines and regulate the power—smoking into the night. When I realized what they were, I backed off: older transformers contain highly toxic PCBs. But nobody seemed to mention that to the workers who, through the evening, poked at the flames. I didn't like it, but there's not much to be said when you're standing in the middle of a scrapyard in a village you've never heard of in a province you've just barely heard of, as the guest of somebody you've just met.[2]

For many years, this was the status quo: a US$200 billion boat nobody would dream of rocking. China produces stuff the west thinks it needs. People in the west consume these goods, usually disposing of them well before their use-by date. Then, these bits and bobs were loaded up and sent back to China on the same ships which brought them over. Far out of sight, people at the bottom of the economic pyramid would take your trash and risk their lives handling it, all hoping to make enough to buy dinner. Yet, most people scrolling comfortably through their iPhones have no idea about this.

But the days of unfettered capitalism, neo-colonialism, and environmental destruction are numbered. In 2018, the Chinese Government passed a national ban on importing foreign trash, including recyclables. This ban had a knock-on effect across the world. Corporations, too, are under an increasingly focused microscope regarding their environmental, social, and governance performance. Stakeholders, not least of which include the general public, are demanding more transparency and accountability. One measure of this is in the number of corporate sustainability reports flooding the market in recent years. Report production is up a hundredfold since 2000.[3] While far from perfect, this is a baby step in the right direction.

All of this begs the question: why? If the proverbial writing's on the wall for bad corporate performance, why hasn't more of the sector changed its ways?

To be clear, the business case for sustainability couldn't be stronger. We know sustainable companies are high-performing companies. According to a report by Merrill Lynch, companies that perform well on environmental, social, and governance metrics outperform their market competitors by up to 3 percent annually.[4] Deloitte reports purpose-driven companies have 40 percent higher employee retention. These companies are also twice as likely to outperform on revenue growth.[5] So what's the problem?

Spoiler alert: it all comes down to money. More specifically, prioritising short-term gain over long-term benefits. No matter what stakeholders demand, most companies are driven primarily by a desire to benefit their shareholders. Executives are measured against how well they perform in this regard. Basically, you're doing well if you're raking in the dough.

Eventually, some of the greedier boards started questioning why they needed to wait until the end of the financial year to see performance. Wouldn't it be better to see things quarter by quarter? That way, they could maximise the potential for profit while minimising the volatility of markets. Now, those in charge of the company were accountable for demonstrating positive financial performance every three months instead of once a year. With that in mind, anything that didn't improve the bottom line quickly wasn't worth considering.

At the bottom of that list is sustainability.

That's because sustainability is one of those things that only pans out in the long run. It's impossible to solve climate change, end world hunger, or even develop a robust employee engagement program in a few years, let alone one quarter. Short-termism won out against having a more long-term outlook, even if this long-term outlook would be more beneficial to an organisation. As long as we can get our payout now, to hell with whether or not sea levels rise or the air becomes unbreathable. Not my problem.

And that's what makes me so hopeful.

Yes, you read that right.

All this pervasive doom and gloom, greed, and money are good. If we know why things aren't changing, we also know what we *need* to change.

I'm starting this book with the corporate sector precisely because of this potential for change. Not just any change, either. Monumental change. Imagine a world where corporations use their power and resources for good. There's no reason you can't marry a profitable company with a positive impact on society. Research shows the two go hand-in-hand. With their access to resources, capital, and influence, we all need to get corporations thinking longer-term once again.

This notion isn't some pie-in-the-sky, Pollyanna fantasy. I've seen firsthand that companies practically turn on a dime when they realise the positive impact of sustainability. My entire career has gone from one board room epiphany to another. Demonstrating the potential billion-dollar effects of green investment on the company and the planet gets people sitting forward in their comfy leather chairs. Defining the most material issues to a business versus a slap-dash approach to strategy perks up the ears of CFOs and operational folks. Explaining to factory managers why happy, healthy workers are more productive has inspired wide-ranging worker betterment programs.

Unfortunately, and it pains me to say this, there are some lost causes. These actors risk negatively capitalising on the positive impact sustainability can bring an organisation. We'll start there with a group I like to call the *Un-Sustainable*. The Un-Sustainable are those specific companies with such a negative impact they can never be considered sustainable. Tobacco, oil, and defence are three of the key industries I'll explore. When your *raison d'être* is predicated entirely on destruction, how can you offset this with a charity fun run, performance report, or slick marketing copy? Every word that comes out of their mouths is pure greenwashing.

Next, I want to cleanse the palette a bit. Having just examined companies on the bad end of the spectrum, what about those on the other end? There are industries and companies which, for a number of reasons, are the most trustworthy out there. Are they perfect? No. But as consumers, these are the players you can take at their word.

Then, I'll explore those industries that may be bad today but can have the same come-to-Jesus moment I've seen many times. That includes aviation, automotive, and some fast-moving consumer goods. Unlike the Un-Sustainable, these industries aren't exactly lost just yet. With a mindset shift, some improved internal mechanisms, and a lot of innovation, they can start to have a truly positive impact.

Lastly, there's an industry so unique it gets its own special section: fast fashion. Why have I pulled them out separately? First, the industry is constantly in the news for its environmental performance. I want to take what often amounts to sensationalist reporting and return it to reality. Fast fashion is also pervasive in our lives. You can't go to any high street and not see an H&M, Zara, or other retailers offering rock-bottom prices for the latest clothes. Given this, we must better understand their history and how they operate. Fast fashion is also very blatant with its greenwashing. That's why it's so important to be able to filter through the BS and reach the truth.

You will be reading quite a few stories highlighting just how bad things have gotten. This is especially true when it comes to corporate greenwashing. These stories might get you down, but please don't lose heart! Throughout this section, keep in mind the notion of change. For as bad as these anecdotes might be, reversed, they could have just as significant a positive impact.

If you're coming at this material for the first time, consider this opportunity a worthwhile (albeit massive) learning curve. For activists and sustainability professionals, use these stories to galvanise and inspire your work. For all of us, we should consider how our own practices, behaviours, and beliefs contribute to the conversation. Are we encouraging more negative behaviours from these corporate actors? Or are we helping realise positive, impactful change?

Chapter 5

The Un-Sustainable

Let's go ahead and get this out of the way.

By their very nature, some industries and companies will never be sustainable. No amount of charity work, photo ops with cute kids, or slick advertising can cover up their insidious business models. They are capital U—Un—Sustainable.

What are my criteria for categorising an industry in this bucket? I take inspiration from a simple methodology championed by engineer Joshua Pearce of Michigan Tech. He notes the '… unwritten rule with industry is you get to make money if you're a benefit to society.'[1] But what happens when you stop being a benefit? What happens when you end up killing more people than you help? As a general rule of thumb, killing off your consumers is a pretty shaky business model. It also makes a company entirely unsustainable right off the line.

In his 2019 study, Pearce comes up with a theoretical foundation and way to quantify whether a business should receive what he calls the 'corporate death penalty'. While he does look at his subjects through the lens of social good, I want to qualify things a bit for our conversation. Now, many industries are terrible for the planet and its people. If one parses the data out far enough, there are very few industries that don't harm society. Don't worry, dear reader. We'll get to each of them in turn.

Who we're going to talk about in this chapter, though, are a special breed. There is a distinct difference between most companies and the Un-Sustainable. Their lack of sustainability isn't simply an unfortunate by-product of growth, like agriculture or fashion. It's not a learned competitive behaviour like in the fishing or fast food industries. Nor is it something R&D, innovation, or consumer pressure can fix. If you want to see that, look at the auto, aviation, and energy sectors.

No. The difference is that the Un-Sustainable can credit their entire existence to death and destruction. I'm not talking philosophically, either. They literally have a business model or product that kills the end-user or causes considerable damage. The only way for these businesses to truly become sustainable is to put themselves out of business.

So, who makes the list? It's pretty short, but I'm sure you'll agree these are the worst of the worst. You can call them the unholy trinity of sustainability: tobacco; defence; and oil.

Tobacco

When I was with Business for Social Responsibility (BSR), we had an unspoken rule guiding the clients we would work with. We'd take them on if a potential client genuinely wanted to change and build back better. That was regardless of their past behaviours and indiscretions. Of course, there were some significant caveats, including engagements with many industries in this chapter. Generally speaking, though, we'd approach clients as if with a fresh slate. It seemed the right thing to do.

I remember the day that changed.

A major global tobacco company came to us to run a human

rights impact assessment. These pretty standard surveys gauge how well a company's operations align with international human rights standards. They were also typically one of an organisation's first steps in its sustainability journey. An assessment includes various sections on elements like forced labour, privacy, and the prevention of torture. It would be reasonably easy to look through a tobacco company's operations and find if anything untoward was happening. If we found something, the assessment's guiding principles explicitly state the activity must cease immediately.

The very first section of the assessment, though, was where we started to question everything. That section supersedes the rest of the assessment because, without it, one cannot have a true adherence to human rights. That section is called 'right to life' and outlines a company's duty to protect life in all its forms. In addition to this, there are also sections on protecting health and wellness. Are you starting to see the issue here?

By their very nature, tobacco companies have the pesky little habit of killing their end-users. That fact immediately put the company at odds with rule #1, the right to life. No matter how well-intended they might have been in pursuing a human rights impact assessment, there was no way in good faith we could conduct one. A great debate raged within our organisation on whether we should look at this as a great way to change one of the world's worst industries. Others thought we would be selling ourselves out and ran the risk of helping them greenwash. Ultimately, we decided against the engagement and stopped taking on tobacco companies as clients.

It was definitely the right move. Unfortunately, it didn't stop that company, or many of its peers, from finding other less ethical consultancies to greenwash for them. One such tobacco company is Imperial Brands, the world's fourth-largest global cigarette producer. Working with Varisk Maplecroft, a British risk management firm, Imperial gauged performance against consumer, worker, supplier, and community risks in their 2016 assessment.[2] In looking at the right to life, the consultancy gave Imperial a low-risk ranking! They even noted the company as having sufficient due diligence standards in place. All these wink, wink, nudge, nudge results do is reward and promulgate bad behaviour.

But Imperial Brands isn't the only tobacco company to undergo such sham assessments. Other notable brands include Philip Morris International, Japan Tobacco, and British American Tobacco. The latter was so proud of their work they tried to make a big PR splash as the first tobacco company to release a human rights impact assessment. Remember what we discussed with 'best of the worst' greenwashing claims. Here is an excellent one, front-and-centre.

What's interesting about all of these assessments is where they focus the reader's attention. Of course, they're not going to put their human rights (lack of) performance in big, bold letters. Unless you know to look at that particular section or ignore the whole thing anyway, you're unlikely to realise they're pulling the proverbial wool. Instead, these assessments focus on supply chain issues like forced labour, child labour, and training. Each of these is table stakes for any company operating today. You can't have kids making your products, no matter your industry. Training is just a good business practice to ensure your operations are streamlined and running at peak performance. Again, something a major multinational is already going to have in place.

Do you remember your time back in high school language classes? If you had the same experience as me, there was always at least one student already native in Spanish, French, or Mandarin because they came from Mexico, Mali, or Macao. What in the world were they doing in the class ... besides just trying to get an easy A? They were the original gaslighters; pretending to be something they weren't.

Essentially, these companies are doing the same thing. They are leading with the glowing reports around stuff they were already doing. No change is happening because no change is necessary. Some, like British Tobacco in their 2020 Human Rights Report,[3] go further. In it, their CEO made aspirational claims of building a 'better tomorrow' and 'reducing the health impact' of their business. Lots of fluff if you ask me.

Beyond human rights impact assessments, tobacco companies think they're the masters of deception. Much of their greenwashing comes in the form of 'look over here, not over there'. One of my favourite examples comes from Natural American Spirit Cigarettes.

Natural American Spirit Cigarettes had a bit of a heyday in the early 2000s. They were trendy among the hipster set because the company claimed it was a healthy, eco-friendly alternative to other cigarettes (as if there were such a thing!). They sat under parent company, Reynolds American; one of the world's largest tobacco manufacturers. While they shut down their Santa Fe headquarters in 2018, their claims are still helpful as a case study.

The company made a number of outlandish claims, including:

'We saved 280,000 paper hand towels in 2010 by installing hand dryers in our Santa Fe office.'

'We saved 30,000 paper cups in 2010 by glazing ceramic mugs for ourselves and for our guests in Santa Fe.'

'Our sales team's hybrid car fleet saved 312 barrels of oil in 2009.'

'We have been 100% wind powered since 2008.'[4]

What makes these claims so egregious is that they focus exclusively on the environmental efforts of the company while completely ignoring the human health tolls their products produce. Remember, sustainability isn't just about the environment. There are so many other elements involved in saving the world. Yet, the company decided to talk about its meagre environmental efforts. Saving paper towels? Come on! While reducing paper use and vehicle emissions is great, Natural American certainly wasn't doing all it could. Being part of such a mega-company would surely mean they could do more than just purchase a few glazed mugs. One must also question where, how, and what they mean by having 100 percent wind-powered production.

Of course, these claims also have the predictable aim of glossing over cigarettes' negative and environmental implications. According to the Campaign for Tobacco-Free Kids, '... cigarette smoke spews more than 7,000 chemicals into the environment, including hundreds that are toxic and at least 69 that cause cancer.'[5] This doesn't even mention the number of cigarette butts thrown away when a smoker finishes their toxic snack. The Campaign notes that people discard at least 5.6 trillion cigarettes every year. As cigarette butts are non-biodegradable, all the wind power in the world won't offset the industry's negative impact.

But hey, at least our guests have lovely mugs to drink their coffee!

Defence

I'm sorry. You won't win any awards for figuring out why defence is top of the list when it comes to unsustainable industries. These companies exist to push and profit from war. The last time I checked, wars are pretty good at killing people. Therefore, we wind up full circle back at the right to life—essential for the sustainability of any business.

When we talk about defence as unsustainable, I'm explicitly limiting my definition of the industry. I do not include aerospace engineering in this section. They'll get their own in just a little while. The real unsustainable part of the defence industry consists of any company that ideates, produces, and sells military equipment. One small exclusion is when that equipment rehabilitates veterans or deals with post-conflict peacebuilding. Never mind the mental gymnastics required to separate those fixing the problem from those who created the problem in the first place. It's better we just focus on the core of the sector.

Selling all the latest gadgets for war to countries worldwide is big business. PwC puts the industry's 2021 revenue performance at nearly US$700 billion.[6] Now if you're thinking this industry is made up of covert arms traders meeting with dictators in gold-gilded rooms, you'd only be partly correct. Most big guns (*groan* ... sorry) are household names. They include the biggest contractor, Lockheed Martin. The company's 2020 revenue was more than US$65 billion, with 89 percent of that coming from defence work.[7] Then there are the likes of Raytheon, Boeing, Northrop, and General Dynamics. These five companies alone account for nearly US$255 billion in revenue.[8]

It's not like the industry is finding more humane ways to conduct warfare or, god forbid, push for peace. A portion of this revenue goes into the research and development of next-generation weaponry. Interestingly, weaponry is the focus when one looks at the sustainability messaging coming out of this sector. Like tobacco, there is no mention of indiscriminate killing of civilians or destruction of infrastructure. Nope. It's all about how green their weapons are.

In a 2021 report titled *The Growing Climate Stakes for the Defense Industry*, Boston Consulting Group looked at how the defence sector could become more carbon friendly. It reads like a surreal Orwellian strategy. The report points out the defence sector can potentially contribute up to 25 percent of total global carbon emissions by 2050.[9] Instead of taking a step back and asking why these tools of war are necessary at all, these high-paid consultants instead looked at how and why to make them greener.

> Contractors that fail to act will face increased pressure from investors and customers, who are placing a higher priority on environmental sustainability in their portfolios. Such companies could incur higher capital costs; they could lose market share or opportunities to supply new products and services that will help defence ministries achieve their nations' climate goals; and they could miss out on public R&D funds related to climate change. Over time, they could also become less cost competitive.[10]

Sustainability and profitability are not strange bedfellows to me. Seeing higher capital costs and decreased competitiveness as a reason for change is totally realistic. I'm not a pacifist. Nor am I dovish when it comes to issues like defence. What I take issue with, however, is the positioning these firms take concerning sustainability. We know you can't be sustainable, so don't try to angle yourselves as such. You can green your operations where possible, but don't think for a minute that makes you an ethical industry. Having the least polluting rocket launchers doesn't negate the fact you still produce rocket launchers.

Over the next century, we'll also see increased resource scarcity. What happens when a country doesn't have enough food or access to water? It starts to fight with its neighbours to get that resource. I can guarantee the defence industry will be there to provide the weapons in those fights. Thus, they exacerbate the very problem they pretend to care about.

Some of the world's biggest defence contractors put out lengthy sustainability reports and even human rights impact assessments. Like big tobacco, the military-industrial complex only discloses what it wants you to know.

- Take Lockheed Martin. Their 2021 Human Rights Report does not mention the 'right to life' central to an impact assessment. Instead, it focuses on ethics, diversity, training, and suppliers. The company's most recent sustainability report is, ironically, titled *Propelled by Principle*.[11]
- French defence contractor, Naval Group, notes developing their sustainability commitment with 'respect to its stakeholders and society at large'. A full 99 percent of their revenue comes from sales of military equipment.[12]
- Also, clocking in at 99 percent of revenue from military equipment is Europe's MBDA. Their CEO, Eric Beranger, pointed out how the company has 'constantly stepped-up our efforts to mitigate our environmental impacts and support

our local communities to offset any negative impacts of our activities'. By the way, MBDA exclusively produces missiles.[13]

Do these companies serve a social good? Until there's global peace, yes.

Do we need all the products defence contractors sell? Perhaps.

Do they deserve to align themselves with the United Nations Sustainable Development Goals and call their operations environmentally friendly? Absolutely not.

Oil

Setting sail from the Canary Islands on March 14, 1967, the 120,000-ton capacity supertanker SS *Torrey Canyon* began its five-day journey north to Wales.[14] By the morning of March 18, the crew spotted the Isles of Scilly off the Cornish coast. That wasn't where they were supposed to be. The ship was off course. The captain was unsure how to navigate the waters without proper mapping of the treacherous area. By the time they found a solution, it was too late. The massive ship ran aground after striking Pollard's Rock, one of the westernmost points in the UK.

Initial efforts at lifting the vessel proved useless. Pressure began to break the ship apart, releasing its destructive contents into the sea. Although oil spills were not unheard of, the size and scale of the Torrey Canyon made this a unique situation. One can only describe what followed as idiotic attempts that exacerbated the problem. Firstly, and following what would have been protocol at the time, fire units used detergent to try and sop up the mess. As we've learned over time, this approach had a deleterious impact on flora and fauna.

Then, the military got involved. On orders from the Prime Minister, the Royal Air Force began to drop bombs on the ship. The idea was to set fire to the spill, thus limiting the extent of the damage.

Unfortunately, high tides and cool temperatures made starting any fire difficult. Instead of regrouping and thinking through other options, the military just bombed the site more. Over twenty-four hours, Gil Crispin notes the military fired a total of 161 1000-lb bombs, 1100 gallons of kerosene, 3000 gallons of Napalm, and sixteen missiles at the ship.

The 1967 grounding of the SS *Torrey Canyon* along the western coast of the UK dumped 36 million gallons of crude oil into the sea. Oil contaminated vast stretches of the Cornish and French coastlines, killing tens of thousands of animals. Toxic first-generation detergents continued to have a negative impact on the area for years following the disaster. Yet, up until that time, nobody was legally responsible for disasters like this. In fact, during the spill itself, the UK undersecretary responsible for handling the disaster said the Government clearly '... has no responsibility in law for what has happened.'[15]

Although it is known to this day as the worst oil spill in UK history, and one of the worst ever, it did bring about the first International Convention on Civil Liability for Oil Pollution Damage. Other international regulations followed, proving to be a bit of a silver lining in an otherwise awful event. Within a decade of the disaster, though, at least three other spills exceeding that of the Torrey Canyon took place. That includes another of history's top ten spills, the Amoco Cadiz, a day before the tenth anniversary of the Torrey Canyon. In fact, the 1970s saw an average of 24.5 large oil spills every year, with 1974–75 the worst on record.[16]

Over the next forty years, we've seen the number of spills decrease. But, their severity has been off the charts. Saddam Hussein purposely initiated the 1991 Gulf War Oil Spill during the Kuwait War, which still holds the record for history's worst spill. There was also the Exxon Valdez in 1989, Fergana Valley in 1992, and BP's Deepwater Horizon in 2010. Deepwater Horizon is still the worst *accidental* oil spill in history.

The last of our unholy trinity, the oil sector is probably the most well-known as being unsustainable. This is most readily apparent in the environmental disasters brought on by tanker spills, most of which were due to human error. We have to remember, too, all that oil was going somewhere. It was to be used for cars, planes, creating plastics, and as a key component in much of our household goods. Oil is so intrinsically intertwined into our daily lives that it's impossible to decouple. Anyone who tells you we can is lying. They're lying just like the oil companies lie to you about how sustainable they are.

Consider the example of ExxonMobil. Formed in 1999 by the merger of Exxon and Mobil, the 2019 Fortune 500 places the American oil and gas giant as the second-largest US corporation by revenue. Given its size, ExxonMobil has immense lobbying power and influence. On paper, they purport to use this for the good of humanity. The reality, though, isn't as altruistic.

Lucky for us, we now have the privilege of hindsight and a ton of informative documents released by the Union of Concerned Scientists (UCS).[17] These reveal just how far the oil giant has gone to steep itself in the language of sustainability while actively working against building a more sustainable future. On the surface, corporate executives say nice things like the '... risk of climate change is real and it warrants action. Ninety percent of emissions come from the consumption of fossil fuels.'[18] This makes you feel as if Exxon is doing its best to decrease its negative impact on the Earth. That is until you witness what they were doing behind the scenes.

The Union's documents highlight how Exxon executives in the 1980s fully understood the negative impact companies in their industry had on the climate. Rather than take this knowledge and do something positive, the company continued publicly denying any links between fossil fuels and climate change for decades. Even more covertly, they funded influential groups that actively discounted the science of climate change. All the while, Exxon publicly called for greater efforts to be made to curb climate change and environmental degradation.

One 1998 memo by the American Petroleum Institute, an industry group that is bankrolled by ExxonMobil and other oil and natural gas companies, laid out a strategy to get the public to 'realise' the 'uncertainties of climate change'. It would target high school science teachers, conduct a media campaign, and distribute 'information kits' that included peer-reviewed papers emphasising 'uncertainty' in climate science.[19]

When this information finally came to light, the public was outraged and pressured ExxonMobil to change its evil ways. Unfortunately, when you're a multi-bajillion dollar corporation, screams from the peons below are hard to hear from the top of your office tower. Sure, they did change a bit of what they were doing, on the surface at least. In 2008, the company publicly stated it would no longer support climate-denial groups.[20] Great! Except they are still part of the American Legislative Exchange Council (ALEC); a group that lobbies US senators and representatives to block action on environmental issues. Not exactly the kind of company you want to keep if you're serious about sustainable practices.

Then, there's the fascinating legal case of Steve Donziger.

A Harvard-educated attorney, one of Donziger's very first cases out of school would become the one to define his entire career.[21] In 1993, he was part of a team working on an environmental case against the oil giant Texaco. For nearly thirty years, Texaco had been dumping millions of gallons of oil throughout the Ecuadorean rainforest in what would become known as the Amazon Chornobyl. Gizmodo notes the total spillage was eighty times worse than that of the BP Deepwater Horizon disaster. Yet, given the victims were primarily poor, rural farmers in the developing world, Texaco could get away with this behaviour unnoticed. The result was a litany of health issues, including cancer, skin conditions, and congenital disabilities.

When the case finally did go to trial in New York, it had ballooned into a 30,000-person class-action lawsuit against Texaco. Shining a light on egregious behaviour wasn't a good look for the company. The need to fix this PR issue became even more apparent when Chevron bought Texaco in 2001. They successfully lobbied to move the case to Ecuador, arguing the system there would be able to provide a fairer environment for a trial. Of course, the real reason was likely because Ecuador does not have jury trials, the system isn't nearly as transparent as in the United States, and media would be less likely to bring their cameras to South America.

Ecuadorian judges saw right through this, ruling in favour of the plaintiffs. The court ordered Chevron to pay US$18 billion in damages to families and to clean up the contaminated areas of the country. Almost immediately (and this is where the real lunacy begins), Chevron appealed the decision. They didn't feel responsible for any wrongdoing by Texaco, even though there was significant corporate precedent on responsibilities of acquiring companies. They argued Texaco had sufficiently cleaned up the toxic waste. Any remaining clean-up was the responsibility of Texaco's local partner, Petroecuador. Chevron promptly moved all its assets out of Ecuador to thwart having to pay any restitution.

The company called a judge from the original trial as a star witness during the appeal trial. The judge claimed Donziger and his team bribed him with half a million dollars to sway the decision. Never mind Donziger had to bootstrap the entire case from the beginning. Suddenly, there was apparently plenty of cash to splash around. Another point of note: the star witness and his family had been moved to the US and were receiving a US$12,000 a month stipend ... from Chevron. Nothing to see here, folks. In a move that should surprise nobody, the judge eventually recanted his claims, saying most of them were fabricated or exaggerated.

Unfortunately for Donziger and the citizens of Ecuador, the die was cast. A 2014 ruling found that even if the star witness was lying, the substance of the case was still the same. It also upheld accusations of fraud against Donziger. In 2018, the International Court of

Arbitration in The Hague further determined widespread fraud in the case and threw out the original compensatory ruling. Chevron also demanded Donziger pay back legal fees to the tune of US$800 million.

But the strange twists and turns of this saga were just beginning.

Before going any further, though, it's important to take a step back (how's that for a cliff-hanger?).

On the surface, this all looks like a pretty stock-standard trial against a major corporation doing evil and trying to get away with it. It definitely is, and these cases are not new. We've mostly all become resigned to the notion that big corporations rarely lose these fights. Their teams of lawyers, and deep pockets, mean they can squash any opposition fairly easily.

What we have with the Ecuadorian case and all the madness that follows is much more sinister. You'll see shortly how this evolved from an environmental lawsuit into a case aimed squarely at discrediting someone's reputation. Nobel laureates, international lawyers, and even the United Nations have all called Chevron's actions beyond the pale and punitive. Even Chevron's legal team has said they were '... going to fight this until hell freezes over, and then we'll fight it on the ice.'[22] By their own numbers, Chevron has already spent over US$1 billion appealing and pursuing this case. Why continue fighting a thirty-year battle that the company already seems to have won?

One of my favourite Chinese idioms is '杀鸡儆猴', which translates to 'kill the chicken to scare the monkey'. This is precisely what's going on here. It's a good old-fashioned scare tactic where Donziger is the sacrificial chicken. In discrediting him, Chevron means to scare all the environmental activist monkeys who would dare take them on in the future. And if you're thinking, what could Chevron possibly do that would scare any environmental lawyer from taking them on ... just read what they've done to Donziger.

We've already seen through claims of witness tampering the beginnings of a character assassination against Donziger. A 2020 release of internal Chevron emails shows a concerted effort to demonise Donziger through a years-long smear campaign that included hiring a private investigator and assembling a legal team numbering in the hundreds. While not exactly on the books, judges sympathetic to Chevron had ruling positions over the various appeals in this case.

One particular judge, US District Judge Lewis A. Kaplan, has been a central character throughout all of this. Kaplan was the judge during the original appeal in 2014. As part of this appeal, he ordered Donziger to turn over his cell phone and laptop to Chevron for further investigation. Donziger refused, citing attorney-client privilege. For this, Kaplan charged him with six counts of contempt. Because of these contempt charges, and based on other rulings by Kaplan, the New York State Bar Association suspended Donziger from practicing law. He has since been disbarred entirely.

When the US Attorney's Office refused to pursue those six criminal contempt charges against Danziger, Kaplan instead appointed a private law firm to prosecute. This move was highly unusual, eliciting concern from legal scholars and senators. What's more concerning is this private firm, meant to impartially represent the US Government, counted Chevron as one of its clients. Both Kaplan and the firm forgot to disclose that bit of information.

It was 2018 by the time Donziger had his day in court. At this stage, he had been fighting Chevron for twenty-five years. In another highly unusual move, Kaplan handpicked the judge for this trial. Usually, selection is random to prevent the type of unethical actions going on here. The judge, Loretta Preska, was hardly impartial. She served on a board for a group receiving substantial contributions from Chevron. Again, this went undisclosed.

Rather than a trial by jury, which Donziger requested, Preska would try the case herself. She quickly disqualified part of his legal team. Then, she ordered Donziger to pay a US$800,000 bond—a US record for this type of case. Once he posted this astronomical sum of money, Judge Preska deemed Donziger a flight risk. In 2019, he was ordered to undergo home detention while awaiting trial. That included an ankle monitor to prevent him from leaving his home.

Soon 2019 turned into 2020, bringing with it the global pandemic. Still, no trial. Donziger would not enter a courtroom again until July 2021, nearly two years since being put on house arrest. During that period, he had no way to earn an income, which didn't matter since he had his bank accounts frozen. Judge Preska found Donziger guilty on all six counts of contempt, sentencing him to the maximum penalty of six months in prison.

In a Covid-safety scheme, Donziger was released from prison and placed back on house arrest. As I write this, he's closing in on 1000 days of detention. A Twitter post by Donziger, who has been posting continuously for the past three years, notes his umpteen days of detention include '... 45 nights in prison, 5 in a halfway house, 890 with an ankle claw, hundreds of harassing phone calls, 10 urine tests, 4 strip searches, [and] major sleep deprivation.'[23] He adds, though, he would '... do it all over for the honor [sic] of representing the indigenous peoples of Ecuador.'

The people of Ecuador still haven't received a single cent. They continue the clean-up.

Chevron posted fourth-quarter 2021 earnings of US$5.1 billion and a record free cash flow of US$21.1 billion.[24] None of their executives, to my knowledge, have been under house arrest.

When I was thinking through this chapter, I hemmed and hawed over including the oil sector as one of the Un-Sustainable. It's not like we can just get rid of them, like tobacco. And there are ways the industry can improve its operations to become greener, unlike the defence industry where that's impossible. But, sticking to the social good definition of a company posited by Pearce, the oil sector certainly kills more than it provides. Because of that, I decided to include it in this chapter.

Beyond this, it's the unsavoury ways the industry uses its power for evil that classifies oil as Un-Sustainable. Exxon and BP are polluting the natural habitats of so much of the planet, seemingly without meaningful consequence. No amount of activism or attention is stopping their expansion, either. A 2022 Guardian investigation[25] found some damning data on just what's in the works. Over the next decade, '… oil and gas projects will produce greenhouse gases equivalent to a decade of CO2 emissions from China.' This will come from no fewer than 195 gigantic projects called carbon bombs. Even worse, the '… dozen biggest oil companies are on track to spend [US] $103m a day for the rest of the decade.'

By looking at BP's sustainability website, you'd assume they're a force for good. They love to use gobbledygook as a way to throw off scrutiny. The site says such things as BP '… will deliver methane measurement by establishing a hierarchy of measurement approaches and have established a new methane intensity target of 0.20 percent by 2025 using that measurement approach.'[26] They also claim sustainability is the foundation of their corporate strategy, and they care about people and the planet.

Then, of course, there's Chevron spending billions of dollars to fight against one man who just wants things to be better. Their sustainability mantra claims, 'Chevron's commitment to sustainability has never been stronger. Our approach is integrated throughout our business to strive to protect the environment, empower people, and get results the right way—today and tomorrow.'[27] That's interesting given everything we know about the Donziger saga. A quick search of the term Ecuador on Chevron's corporate website brings one to a list of media releases that come off as just a bit defensive.[28]

> 'Chevron is defending itself against false allegations that it is responsible for alleged environmental and social harms in the Amazon region of Ecuador.'
>
> 'For years, the Ecuadorian government, activist groups and calculating lawyers have used images of oil pits, spills and indigenous people in the Amazon to mislead the public.'
>
> 'Chevron has never operated in Ecuador.'

Cool.

There's no need to even consider the tell-tale signs of greenwashing here. When companies from any of these sectors come to you with how green they are, kick them to the curb. Unless, of course, they're making a big splash about shutting down shop and going home. Until that time, you should really pay them no mind.

In that spirit, neither will I.

Chapter 6

Those You Can Trust

Think about the last time you were at the airport.

Maybe you were headed off to an important business meeting to secure a promotion-clinching deal. Perhaps it was to see relatives for a family reunion. Or, you might have been skipping town for a long weekend away with your partner. You get your boarding pass, go through security, and patiently wait at the gate for the agent to call your flight. Even with all the screaming children, flustered latecomers, and scratchy announcements on the intercom, it's hard to contain your excitement. Who doesn't love to jet off somewhere new?

Now you're on the gangway, inching along the carpeted tunnel surrounded by witty advertisements from some of the world's big banks. The air's crisp, the light's bright, and the smell is one of a kind. The very friendly flight attendant smiles and checks your ticket. You're in! Dodging the person who keeps trying to fit their carry-on in an overhead bin three times too small, the couple arguing over the aisle or middle seat, and someone very brazenly pushing against the flow of passengers, you take your seat. Buckled up with a book in hand, the plane speeds down the runway and shoots up into the sky. Your eyes gently grow heavy as the hum of the engines lulls you to sleep.

From the moment you pack your bags to when you arrive at your destination, there are many things to worry about. But unless you're heading to a conflict zone, those things are usually pretty pedestrian: wallet; passport; change of underwear. The one thing most of us don't worry about, let alone consciously think of, is whether or not the people running this show can get us from A to B. Even when the pilot's voice comes on over the loudspeaker, few people would immediately wonder if the team in the cockpit can do their job. They're so out of sight and out of mind I'd bet some of you couldn't care less. How many of us still pay attention to the safety demonstrations?

Which is odd, right? These people, who you've never met and have no information about, literally have your life in their hands. Yet we take a *laissez-faire* attitude to the whole situation. Why? Trust.

You trust they have the hours and hours of training to fly this bird and land her.

You trust the people who serve you drinks can also save your life.

You trust there are measures in place should anything go wrong.

Trust is a crazy thing. Scientists and psychologists have studied it for decades. Charles H. Green even developed what he's termed the Trust Equation.[1] This Equation contains the four major components of trust and measures a Trust Quotient, much like an intelligence or emotional quotient. You can even take a free quiz online to find out your TQ.[2] For those wondering, mine is 6.5, just below the average score of 7. It seems I need to work a bit on my self-orientation!

Green argues trust comes from having a balance of four things: credibility; reliability; intimacy; and, self-orientation. The first three are elements that help boost someone's trust in you. Credibility is whether or not one has the experience, knowledge, and expertise to do something. Reliability refers to their track record. Not only can they deliver, but have they delivered consistently? Intimacy is all about feeling secure. Can I be vulnerable with you, even with my insecurities and fears? The fourth element, self-orientation, focuses on how egoistic a person might be. Do their interests align with mine, or are they just in this for themselves? The higher the self-orientation, the more significant the negative impact on TQ.

So if we go back to our innate trust in a flight crew, does it mesh with Green's equation? To get a pilot's license in much of the world requires at least 250 hours of supervised flying time (not to be outdone, the US requires 1500 hours).[3] This comes with lengthy classroom training and hundreds of hours of apprentice flying. So, yeah, I think this makes them credible. Are they reliable? It's rare to see catastrophic flight failures, especially given how many flights take off daily. We always hear about how safe flying is versus driving, and the latest stats back this up. A Harvard study[4] found your odds of dying in a plane crash are around 1 in 11 million. A car crash is 1 in 5000. So you're more likely to become president of the United States than ever die while flying. And, finally, we've already covered just how intimate a bond we have with these folks who constitute the flight crew.

What about self-orientation? I highly doubt people become pilots out of some sense of ego. Sure, they may have been aces back in the military. But once they start flying commercial, what's the incentive for them to be overly self-oriented? Ultimately they're trying to get to their destination safely, just like we are. The only difference is they're in command.

Combining these factors, it's no wonder we trust flight crews with our lives.

But it's not just interpersonal relationships where trust is essential. As consumers, we also put a certain amount of faith in the companies we support and the products we buy. The big players know this and play around with our impressions of them. In essence, many are trying to game Green's Trust Quotient. In the marketing world, we call these the emotional and functional benefits that consumers believe a company possesses.

Graham Robertson, author of *Beloved Brands*, is one of the premier gurus on how companies position themselves for trust. I had the pleasure of working with him a bit during my time in Shanghai. He talks a lot about how brands can truly differentiate themselves in a highly competitive market, finding their winning zones versus peers. Part of this involves identifying a brand's core strength or unique selling proposition. These sit at the bottom of the consumer benefits ladder, along with what a product does and what consumers want. Where the selling really starts, and trust begins, is in how a brand positions this product versus what consumers receive (the functional benefits) and how this makes them feel (the emotional benefits).

What does all this jargon mean? Functional benefits could be how a product simplifies your life, makes you healthier, or saves you money. Emotional benefits include getting noticed, being liked, and sating curiosity. HotSpex, voted North America's most innovative market research firm, has taken a lot of this and amped it up on steroids. Over the past decade, they've gathered the top 5000 emotional drivers and '... mathematically and scientifically reduced to 88 that predict behaviour with 92 percent accuracy.'[5] With these, they can empower brands to build out personas, associations, and values for any consumer. Tossing these benefits into a magic marketing machine can help a company build consumer trust critical to success.

All of this reveals a few things. First, brands put a lot of time, effort, and money into presenting themselves to you in the best light. Second, consumers are so predictable we can be reduced to less than one hundred emotions that motivate us. That means we're a commodity that's easy to manipulate. Third, this throws the Trust Quotient out the window unless we know where to turn for authenticity.

When it comes to greenwashing, this is our ultimate question.

As we saw in the last chapter, it's sometimes straightforward to determine if a company is being authentic. I doubt any of you would have been ignorant reading about some of the greenwashing happening in the oil and extractives sectors. We don't usually associate businesses creating missiles, guns, and drones with being good corporate citizens. To put this in the parlance of the Trust Quotient, companies in this category would score very high on self-orientation. It doesn't get more selfish than killing your customer base, right?

What it means to be trustworthy can also change over time. You only have to go back to the 1940s and '50s to see doctors enthusiastically endorsing tobacco products. As we wade through this current era of transition, where consumers are demanding more of the private sector, you can expect messaging to evolve as well. Companies considered authentic and trustworthy today may not be in the not-too-distant future. So how are you supposed to identify what's authentic? That's not just for today when you go to the shops, but also a year, five years, a decade from now. Which products and brands can you immediately turn to, comforted in the knowledge they won't be greenwashing the hell out of what they're selling?

At this stage, I know most of you are probably thinking, *Where are all the stories about evil companies lying to us about how good they are? I want to find out more reasons to hate them!* Close to a hundred pages in, we've certainly covered a bit of that. But it's only fair to give credit where credit's due, right? That's why I thought to include a few words on recognising these companies you can truly trust.

To be clear, there's no such thing as a perfect company. Even the most mission-driven, socially responsible, ethical B-Corps will still have some problems when you look under the hood. That's because our era of transition encompasses more than just how a company presents itself. It's also about their operations, approaches, and ways of thinking. In fact, the desire to be a better corporate citizen is never-ending. As society evolves, companies need to evolve alongside it. While most activists think companies can flip a switch and change for the better, all that change will take time. I liken it to a massive battleship. To turn the thing around on a dime is unrealistic. It's the same with corporate behaviour, practices, and social responsibility. We just have to be patient.

I'm not ignorant that patience and time are two finite resources we're clearly running low on. Add to that sanity (because how can you be sane if you're actually paying attention?) and it's no wonder people want such immediate change. My argument, though, is that there are far more companies trying to do better than those who would put profit before planet. I get that's a pretty rich statement since I predicate much of this book on the lies companies spin. But for every Un-Sustainable company in the oil, defence, or tobacco industries, there are heaps more working to improve what they do.

This is easily recognised if we look at the world's major industries. While not exactly exhaustive, there are about twenty-two big industry buckets most of us see daily.[6] Those range from consumer-packaged goods, like what we find on the grocery shelf, to tech, travel, and construction. Of these twenty-two industries, there are only four I would consider being Un-Sustainable. We've just gone through three of these in the previous chapter. The only additional culprit is the gold and mining sector. As much as I'd love to include every single industry, I doubt very much you want to read a thousand-page treatise on greenwashing.

Out of the twenty-two, I would deem three mostly trustworthy today. These are the pharmaceutical, medical technology, and grocery sectors. Again, are they perfect? Absolutely not. But if we look at them broadly, we can see they all have one thing in common. Each is part of a highly regulated sector. Companies that are highly regulated are typically the ones who come around to change the fastest—that's because the government says they have to. I'm a big fan of government command and control when it comes to moving the needle on sustainability. They don't have to do everything, but a little nudge in the right direction never hurts.

Of course, not every highly regulated industry is going to be a sustainable one. Defence, oil, and tobacco are all highly regulated, but for various reasons don't make the cut. That comes down, again, to their self-orientation but also their size. At this stage, these Un-

Sustainable sectors have become so big they're almost impossible to truly regulate.

It's interesting when we juxtapose these three trustworthy sectors against Green's Trust Quotient. Given government regulation, each industry has an intrinsic level of credibility. Are they reliable? As entire industries, yes. Sure, individual companies may have problems with stock, recalls, or unsavoury behaviour. Taken in total, though, reliability is there. We also see a high level of intimacy. Pharma and med tech literally save our lives, while grocery stores (not the individual brands they stock) are part of our daily experience. A recent poll on the most trusted brands in Australia even put Woolworths grocery stores in the top spot.[7] So if you want to start narrowing down those industries and companies least likely to be lying about their sustainable endeavours, keep the highly regulated in mind.

That's all well and good for identifying trustworthy companies today, I hear you screaming. Remember, this whole book is just a snapshot in time. I have no crystal ball to predict the future, and things are changing so fast. But as things continue to shift, morph, and evolve, what rules of thumb should be kept in mind? I hope it's not too much of a bastardisation (sorry, Mr Green!), but let's keep with the TQ framing I've been shoving down your throats.

The first thing to look at is a company's experience with sustainability: its credibility. As I'll mention over and over, companies start their sustainability journey at different times in different ways. You can't expect a start-up restaurant to have the same environmental, social and governance (ESG) processes as a century-old multinational. It's not like for like. Instead, you have to judge performance on what a company knows it should be doing and how they're doing it.

China provides another great anecdote to this point. Being the 'world's factory', companies operating in China have had almost forty years of deep experience perfecting supply chains, logistics, and manufacturing. In turn, these were also the first areas where we saw ESG improvements over a decade ago. Now, it's rare to find a multinational with supply chains running through China that doesn't have its house in order. They have experience, and thus credibility, in this area.

On the other hand, labour rights are a relatively new area of focus. Of course, this doesn't give any company a pass for having a bad record. But one can't expect labour rights to operate at the same level as supply chain performance. They don't have the years of experience under their belt, and as we know, the transition takes time. We'll see this very point in the fast fashion chapter. They may be able to take credit for swift product turnaround, but there's no way we should take their word on labour rights as credible.

Next, consider a company's track record: its reliability. Do they continue to lie, cheat, and steal? Or are they making incremental improvements in a positive direction? RepTrak, a research company with the world's largest reputation benchmarking database, annually reports on the world's top hundred most reputable brands.[8] Scoring is based on consumer perceptions of how a brand innovates and handles crises, as well as if they have a great product or service. They also included ESG aspects like corporate citizenship, diversity, and work standards as part of the response methodology. This doesn't, in and of itself, make a brand sustainable. But the list does serve as a good proxy for how consumers perceive a brand's reliability.

Of the top ten brands, seven have a very established history. We're talking Rolex, Ferrari, and Lego, to name a few. What's clear from this is consumers associate age and pedigree with reliability, which makes sense. Unreliable brands typically don't last very long. Therefore, the longer you've been around, the more reliable you must be. This also holds true with sustainability, where time in practice should ideally lead to a stronger track record.

Third, find out why a company became sustainable in the first place: this can demonstrate intimacy. As much as I'd love to say companies change because they care, they will mostly only make positive changes when they absolutely have to. Sometimes this is because of a public relations disaster. Nike didn't care much about their supplier factories in Asia until the world discovered children were working in sweatshops making running shoes.[9] Other times it comes down to public pressure. We're seeing this today in how financial institutions are paying closer attention to their portfolio companies, increasingly turning away planet-killing investments in oil and gas. Still others, rightly so, see moves towards being more sustainable as a way to differentiate and attract new customers. If you want to identify companies that might be especially cautious about greenwashing, look to those making the news for all the wrong reasons. With such exposure comes honesty, intimacy, and hopefully a bit of humility. Yet while it's always interesting to find out *why* a company begins to improve, the 'why' isn't nearly as important as the 'how' of putting in the work to change. I strongly believe the ends justify the means. Just use a company's sustainability origin story as another piece of data.

And lastly, you've got to ask yourself what a company is actually going to get out of greenwashing: their self-orientation. This is one of the significant differences between highly regulated industries like pharmaceuticals and, say, oil. At the end of the day, remember that greenwashing is a way for companies to appear more sustainable so they can sell more products and/or cover up some pretty unscrupulous behaviour. We know all the ways oil might benefit from greenwashing. But what about pharma? By its very nature, pharma is meant to do good. Barring any conspiracy theories about Bill Gates and microchips, they work to develop products that heal the sick and advance society. While governments regulate them, they also provide the best of the bunch with grants and funding to continue research. Plus, the products sell themselves once they hit the shelf. So, what's the point of risking all this to pretend to be something they're not?

So, mathematicians, how many industries does that leave? That's right, fifteen industries that are neither mostly trustworthy nor entirely untrustworthy. These are industries I would classify as broadly improving.

This improvement is highly matrixed. It depends on factors like the size of a business, how long they've been working on improving, and who's leading the company at any given time. While it might be disappointing to read that there are only a very small handful of companies worthy of trust, know this is improving day by day. To paraphrase Martin Luther King Jr., the arc of history always bends towards justice. And it's towards these companies taking great strides to improve that we now turn.

Chapter 7

Those That Can Change

Here's my take.

Three prominent groups can significantly impact building a more sustainable future: individuals; governments; and companies. Since the beginning of the modern environmental movement in the late 1950s, we've relied almost exclusively on bottom-up people power. The overused Gandhi quote about being the change we want to see in the world sums this up perfectly. And for a time, this worked quite well. Rachel Carson was able to drive monumental changes in the use of harmful pesticides through her work, *Silent Spring*. Activists began to collaborate and develop the foundations for organisations like Greenpeace, the National Resources Defence Council, and the World Wildlife Fund.

Although household names today, it took a concerted effort to get these off the ground. Hell—even politicians were playing their part. A group of American senators, including future Secretary of State John Kerry, held the first Earth Day in April 1972. This period, what I like to call Sustainability 1.0, was one characterised by hope and where people embraced the need for a new way forward. It certainly looked like we were all working together and collaboratively to ensure a brighter future.

But that's only been able to get us so far. In the beginning, and before we had to deal with the scale of environmental collapse we're seeing today, a very vocal minority seemed appropriate. Today, the issues are far more significant than any group of individuals can handle. So the apparent actor to step in would be the government, right? As I laid out in my first book, governments were in an even worse position to do anything about sustainability than individuals because they didn't have the tools, know-how, or historical precedent to act. As examined in Chapter 5, the 1968 grounding of the SS *Torrey Canyon* along the western coast of the UK dumped 36 million gallons of crude oil into the sea—it is known to this day as the worst oil spill in UK history. And yet, the UK Undersecretary claimed the Government clearly '... has no responsibility in law for what has happened.'[1] Although the spill eventually led to the creation of the first International Convention on Civil Liability for Oil Pollution Damage, it's not like governments have done a huge amount to build a better future since that time. There are few, if any, on track to meet their Paris climate targets. Others profit off oil and other destructive elements. That, more than anything else, should show you just how much governments care about sustainability.

So, individual people cannot combat our shared problems because they lack the numbers. Governments can't because they lack a general interest in doing anything. That leaves us with corporations: the private sector.

In my twenty years of experience around the world, I've converted to this idea that the private sector will be the ones to get us out of this mess. In all fairness, they've been responsible for a lot of it, so it's on them to fix. But beyond this, the private sector has a few things both individuals and governments do not. For one, they have an interest in building a better future. That's because there is a business imperative to being sustainable. More sustainable companies outperform their non-sustainable peers in several metrics. Those that rank well on ESG metrics have outperformed the market by up to 3 percent per year over the past five years, according to Merrill.[2] According to the *New York Times*, employees at purpose-driven companies are 1.4 times more engaged and enjoy 1.7 times more job satisfaction.[3] These employees are also three times more likely to stay at their company.

Consumers are also demanding more from the companies they patronise. In their most recent sustainability tracker,[4] A.C. Nielsen show 81 percent of consumers say it's extremely or very important for companies to implement programs to improve the environment. They note 66 percent of general consumers, and 73 percent of millennials, were willing to pay more for sustainable goods. Consumers in the developing world, those most at risk from climate change, are also the most favourable to sustainable practices. There is now an expectation for brands to be ethical, with 90 percent of consumers in Asia-Pacific wanting brands to stand for something. Beyond this, a recent GlobalData report[5] shows 'almost 38% Australians, 52% Chinese and 56% Indians' are influenced by the environmental, ethical, or social responsibility records of a company.

Beyond sustainability making pure business sense, the private sector also has access to capital and resources that governments and individuals could only dream of. Since they are part and parcel of our daily lives, private sector companies have grown to overshadow some national economies. *Business Insider* ran a fun little column back in 2018 listing the size of companies compared to entire countries.[6] You had those on the smaller end of the scale, like Spotify outpacing the entire GDP of Mauritania or Netflix that of Malta. Then you had the big boys. In 2017, Apple's revenues outpaced Portugal's GDP. German automobile maker, Volkswagen, was on par with the GDP of Chile in 2016. Of course, sitting at the top of the list was Walmart, which would be the twenty-fourth largest country in the world, and sixth largest in the Eurozone, based on GDP.

If these companies, and countless others, are the size of entire nations, then can't they also have as significant an impact? Given how they operate, without burdensome bureaucracy and red tape, I'd argue they could actually get more done if put to the task. Plus, they have entire cadres of employees at the ready to make change. The big question is whether or not private sector companies are willing to innovate and change their own approaches to build a more sustainable future.

I'm happy to report most are.

But this isn't a new thing. The private sector has been building more sustainable business models, creating positive societal impact, and shoring up its governance structures for well over a decade. Companies have been ramping up their research and development processes to meet demand in response to consumer pressure. The Carbon Disclosure Project (CDP) ranked some of the world's biggest brands in terms of how ready they are to respond to consumer needs around sustainability. Coming out on top were Nestle, Unilever, and L'Oréal. L'Oréal, for example, is actively innovating to replace petrochemicals with natural, biodegradable ingredients. Unilever was the first among four major FMCG companies to have developed vegan personal care product ranges.

Most consumers, though, still view the private sector as the enemy of progress. Sure, some companies are absolutely vile and we should treat them as combatants. But by and large, there is positive change happening. As I mentioned in the previous chapter, this change will take time—if we let companies continue their journey of progress, our patience will be rewarded with a world vastly different from the one we have today.

Instead of just badgering the private sector with all the bad things they've been doing, I also wanted to share some of my personal experiences with and knowledge of companies making every effort to improve. This isn't to absolve them of any greenwashing sins they've committed. It's also not meant to give them a pass on any unscrupulous behaviours of the past, present, or future. Instead, my goal is to show you how a little bit of innovative thinking can move a company from being unsustainable and detrimental to humanity to one that can do far more than any individual or government.

There are so many examples of companies changing for the better I had a hard time thinking of the best way to organise these juicy tidbits of information. Ultimately, I went back to basics. Now I know you've just had a history lesson. I will also have to subject you to a bit more sustainability education, so bear with me, please. It'll all be worth it in the end.

   When you hear the word sustainability, what immediately comes to mind? If you're like most people, it's likely the environment. While hugging the trees is a critical part of sustainability, it's not the only thing we're talking about. Overall, sustainability seems to be the most confusing thing about being, well, sustainable. That's because it serves as an umbrella term for many topic areas in the cause-centric world. It's multifaceted and prone to different definitions. Even daily, I use the word sustainability interchangeably with several other terms. It's a great catch-all, but that can lead to confusion.

To get over that confusion, let's break the term down into its individual parts.

Traditionally, sustainability consists of three distinct pillars: environment; society; and economy. Some people toss in culture as another pillar, which is fine. Because each pillar is so broad, it can consist of many issues. The environment can include ecology, water stewardship, air pollution, environmentalism, and so on. The societal pillar is chiefly concerned with human rights, corporate responsibility, and community affairs. Economy deals with matters like the private sector, supply chains, and sustainable finance. I could continue for pages, but I won't. Just understand that everything related to saving the world sits under sustainability.

Sustainability may have a particular context depending on where you live, your understanding of the field, and your line of work. In China, new on its national sustainability journey, the word most often implies corporate responsibility. For countries of the European Union, sustainability has to do with regulation and governance. Left-leaning places like Melbourne or San Francisco would use sustainability interchangeably with environmentalism. Confusing? I know. Yet all of these would be correct.

To wrangle the case studies I want to present into something manageable, I decided to go with the acronym ESG: environment; social; governance. The environmental aspects of ESG relate to a company's negative or positive impact on the Earth. This could be through its air or water emissions if it's a factory, green building record if it's an architecture firm, or carbon offsets if it's an executive jet-setting around the world.

The social aspects of ESG deal with how a company treats its workers and stakeholders in the local community. Are all labour rights and employment protections in place? Do workers receive training or education, and are they offered opportunities to grow? Is the company a good corporate citizen, and do they act responsibly towards their local communities? Do they support the local economy or, like Walmart and Starbucks, aim to put mom-and-pop shops out of business?

Lastly, governance involves how a business monitors itself. Good governance models allow for transparency and the full disclosure of information. Even better models tie executive and board remuneration to ESG KPIs. This term only has to do with politics if you're talking about the frustrating internal politics of an organisation.

You'll see ESG indicators in corporate responsibility reports, materiality analyses, and financial statements. Investors especially love ESG as a way to value companies they may or may not invest in. Major multinational organisations use it to decide which factories and suppliers to work with since supplier audits generally include this information. If you're a factory in Dhaka and you get a bad ESG audit, good luck getting any company in the United States or Europe to work with you. That's why this little acronym is so crucial to the field of sustainability.

And that's why it will guide us through the rest of this section.

Over the following pages, I'm going to introduce you to some of the best cases I can think of when it comes to companies using innovation to change the world for the better. These cases come from all parts of the private sector: shipping to forestry; consumer goods to technology; automotive to that lovely cup of tea or coffee you're sipping on right now. Hopefully, this will serve as a palette cleanser from all the muck I've been forcing down your throat as we explored greenwashing. If not, they'll at least give you fabulous little factoids to make you look good at your next cocktail party.

Environment

You know that famous sketch showing the evolution of humans?

On the left is our not-so-distant forebear, the ape, crouched down as if he needed to grapple with a global pandemic, financial meltdown, and collapse of society. Then there are progressive images as the ape starts to stand erect, acquire tools, and lose a whole lot of hair. By the time we get to the rightmost image, the only thing missing is a Hugo Boss suit and an iPhone. I'm sure in a thousand years, the sketch will continue, but we'll be *de*volving instead.

This is much like the well-trod evolutionary trajectory private sector companies set themselves on when they start their sustainability journey. The first stop on this journey is some philanthropic or charitable undertaking. Usually, this is a passion project of a senior executive, a fun run for a local cause, or volunteering during the holidays. Inevitably, the next thing this company will look to is environmental stewardship. We're not talking about installing next-generation effluent disposal systems just yet. It's more like starting a recycling program or encouraging riding bicycles to the office.

While some companies will continue to evolve, many put a lot of emphasis on this environmental stage of the journey. This, along with all the activist focus on environmentalism, is why most people associate sustainability with 'green'. It's the loudest, most in-your-face aspect of the practice. As we've seen, though, this is only a tiny part of the story. Yet, it's also where there are a lot of great case studies to choose from.

If you've been paying attention, you'll also know it's where we find many greenwashing cases. You'd be right. So to get over the hurdle of being trustworthy cases, I've decided on a few where I have personal experience working with these companies. This has the added benefit of giving you a behind-the-scenes look at what they've done and how they've done it. In the interest of full transparency, I also want to note these are in no way sponsored posts. I mean ... how could I write a book all about greenwashing and then do something unethical like that? These are just some of the cases I find most interesting based on the hundreds I've worked on over my career.

My career has taken me in a lot of different directions. Where I find myself gravitating to the most is an area of sustainability that's proven one of the most significant barriers to overcome: effective communications. We're terrible at it. It's either fluffity-fluff-fluff or dystopian doom-and-gloom. Corporations are no better. Unfortunately, the squeaky wheels are the ones that get the oil. But the share of voice isn't directly correlated in any way, shape, or form to actual impact. If anything, it's usually the opposite.

Which begs the question: why aren't companies that are actually having a positive impact making more of a splash? I think it all comes down to intention. For companies that really care, it might seem disingenuous to boast about all the good things you're doing. If you're making a positive impact and have the planet's best interests at heart, then why not let your work speak for itself? Although a nice altruistic approach, it's certainly not the most pragmatic if we want more companies to join the fight.

A case in point is an organisation revolutionising how we do business: DocuSign. If you work at any company with more than three people, you've probably used DocuSign before. DocuSign created digital platforms—for those reading this book from the distant future—to house corporate documentation and streamline approval processes. They've taken all the paperwork that used to sit in millions of filing cabinets worldwide and moved everything online.

DocuSign approached me about a year ago to help them research the state of greenwashing. Sure, I thought, this would be great timing given I was knee-deep in the subject already. But my Spidey-senses started to tingle. Would this be an exercise in greenwashing to validate a Silicon Valley tech giant jumping on the sustainability bandwagon?

Then they dropped a bombshell I had never considered. Like most businesses having a significant impact, they've been doing heaps of work under the radar. DocuSign has not only changed how we do business but also had a massive environmental impact through its work.

Think about it. Before the platform, most HR teams and businesses did their work on paper. Businesses loved paper so much there were two hit comedy shows, one on either side of the Atlantic, about a paper company! At the risk of sounding like an advertisement, DocuSign eliminated the need for all that paper. They've put Dunder Mifflin out of business and saved millions of trees in the process.

> The paper industry is the fifth largest consumer of energy in the world. Producing one tonne of paper requires 1,440 litres of oil, 2.3 cubic metres of landfill space, 4,000 kilowatts of energy, and 26,500 litres of water. To make just one piece of A4 paper, 10 litres of water goes down the drain. Every year, paper usage grows by 22% for an average business, yet nearly half of it gets thrown away daily.[7]

If so much goes into making a single piece of paper, what's DocuSign's impact in changing all this? In Asia-Pacific alone, the platform saves 14 million kilograms of wood and 350 million litres of water annually. It also keeps 33 million kilograms of CO_2 from getting belched into the atmosphere, the same as removing 7000 cars from highways.

Globally, they've saved over 2.5 million trees and 2.5 billion gallons of water.[8] That's a lot to crow about.

What about innovation with an even more considerable potential for impact? Electric cars have been a hot topic for years. As a kid growing up in Southern California, the news ran a piece on them weekly. One that sticks in my mind was the announcement all cars would be electric vehicles (EVs) by 2010. That would have been around 1990 or so. Clearly we missed that target, but the transition from fossil fuel vehicles continues.

What's taking so long? I've already reiterated that innovation doesn't happen overnight, especially in sustainability. It takes a lot of planning, research, and money. But companies leading our transition to a more sustainable world are willing to undertake the challenge and invest where needed. One such company is BMW. They're publicly working to electrify their fleet of vehicles, making them pioneers in the automotive space. They aim to have at least one electric vehicle within each car line by 2030. That means at least half of all new vehicles will be electric. This isn't just fluff, either. I had the opportunity last year to attend the launch of two new EVs in Singapore, which means the cars are actually available and not some figment of BMW's imagination. They are also investing in the infrastructure necessary to run EVs, like charging ports, for the entire Malay Peninsula.

BMW has also pledged to cut in half CO_2 emissions per vehicle by 2030 and reduce their cars' total environmental impact, from production through utilisation, by 40 percent. In speaking with BMW's Vice President of Sustainability, Dr Thomas Becker, he added that none of this would happen using carbon offsets. Offsets are, in my opinion, just a fancy way of deflecting and assuaging guilt for companies that don't really want to change. BMW's moves are the long-term, sustainable change we need to see companies making.

True to their business model, BMW is also doubling down on emphasising luxury as part of their offering. Instead of just not talking about going green, the company is working to change consumer perceptions of what green luxury means. This is a potentially major shift from the ongoing association of sustainability with fringe activists, living off the grid, and basically going back to the Stone Age. Instead, BMW promotes the vehicles' circular design as the 'purest form of Sheer Driving Pleasure'.[9] The company has realised there's no reason luxury and sustainability have to be strange bedfellows. Through this, they're addressing consumer demands for greener products while shifting the paradigm within their industry.

One final area I find super fascinating is how far and wide the technology sector impacts building a more sustainable future. Even before Elon wowed us with his Teslas and lied to us about his Hyperloop, private sector companies had adopted new technologies towards good. Nowhere is this on better display than in China. During the ten years I lived and worked there, I witnessed firsthand the dramatic rise of technologies to address myriad environmental, social, and governance issues.

One final area I find super fascinating is how far and wide the technology sector impacts building a more sustainable future. Even before Elon wowed us with his Teslas and lied to us about his Hyperloop, private sector companies had adopted new technologies towards good. Nowhere is this on better display than in China. During the ten years I lived and worked there, I witnessed firsthand the dramatic rise of technologies to address myriad environmental, social, and governance issues.

Environmental concerns, for example, are front-and-centre throughout China. People step outside and choke on pollution. They walk past dirty waterways. Many eat tainted food. A perfect storm emerged where governmental, business, and individual priorities aligned to fix these issues. Digital technology stepped in to monitor, track, and name and shame the big polluters. I distinctly remember one winter morning when the pollution was so bad you couldn't see the futuristic skyscrapers across the Huangpu River. That's when real-time air quality monitoring apps began to show up. It seemed overnight that people downloaded these trackers and became mini-meteorologists. You'd chat with friends about the levels of PM10 versus PM2.5 particulates in the atmosphere and whether or not to wear a facemask outside. Alongside air quality monitoring apps are real-time data trackers for factory emissions, individual carbon footprints, and wastewater and effluent discharge. Some of the most prominent players in this space are the Institute of Environment and Public Affairs, IQAir, and Swarovski, respectively. In a virtuous cycle, this exposure to information forced the Government to act. The latest stats indicate China's reduced its air pollution nearly as much in seven years as the US did in three decades.[10] Without technology's pivotal role, I'm not sure this would have ever happened.

Social

Like the runways of Paris, New York, and Milan, sustainability goes through seasons. What's a hot focus today may not be tomorrow. For a long time, the big thing was corporate social responsibility, or CSR. Companies clamoured to outdo one another when it came to being good corporate citizens. Some hosted large-scale charity drives. Others sent hordes of volunteers to do beach clean-ups. If there was a photo-op with some kids, watch out! Companies could smell that a mile away.

As companies continued to evolve beyond just the first few stages of sustainability, other (arguably more meaningful) areas of sustainability replaced CSR. We already mentioned the early phases dealing with philanthropy, which is followed closely by work in environmentalism. Towards the middle of the evolutionary trajectory, companies will likely address their internal operations, fix supply chain issues, streamline logistics, or treat their employees better.

The pendulum is now swinging the other way and social sustainability is back in vogue. This time, though, companies are taking what was once essentially charity and making it broader, driven by stakeholder needs, and ultimately more impactful.

I often cite an example of excellence in social sustainability from a company people love to hate: Walmart. For years, do-gooders have fought against Walmart and its hegemonic family for their treatment of workers both stateside and overseas. They cite Walmart's propensity to purposely squash mom-and-pop stores worldwide in a quest for domination. Walmart also has an unsavoury track record of employing the undereducated and then paying them slave wages without benefits.

If people keep demanding low, low prices, though, someone has to pay. While grinding down the wages of American employees might do a bit to keep prices low, where cost savings truly materialise is on the production side of the equation. Where does that production happen? Contrary to the aw-shucks, all-American company image, it's an open secret most of Walmart's production is outside the United States. One of their biggest production centres is China, where factories buzz at breakneck speed to produce products you think you need.

There is plenty of blame to pass around. As a corporate operator and top of the Fortune rankings, Walmart certainly doesn't have its hands clean. They employed production at the Rana garment factory, which we'll learn more about shortly. Many other locations continue to have piss-poor working conditions. They underpay their suppliers, if they pay them at all. But let's take a trip to the southern Chinese city of Shenzhen, just across the border with Hong Kong. It's here Walmart has its China headquarters and where most of the operations for its nearly one hundred mainland China factories happens. It's quite the operation to manage. Thousands of staff ensure things run as they should, without delays or hiccoughs. If production goes wrong here, you're not getting that doll in time for Christmas.

It's not surprising Walmart knows what it's doing with production. If not, how did it become the largest consumer company in the world? What you might find surprising are the conditions in which all this happens. Unless you've visited a factory in China, or across much of South-East Asia, it's hard to conceptualise what goes on inside. Sure, we have our preconceived notions of what a factory looks like. Hundreds of tired workers huddle in cold rooms, at risk of a missing hand if they doze off next to a machine. Dormitories have no air conditioning in the middle of summer and little heating when it's freezing outside. Automated machines buzz quietly while specialised technicians in white coats keep things rolling along in near hospital-like conditions.

Huh?

That last part might have thrown you a bit, but it's absolutely true. In China's quest for development, it is adopting automation at record levels. This adoption then trickles down to the factory level, where companies like Walmart invest and upgrade their operations. Whenever I go into a factory, I'm amazed at how safe and technologically advanced the conditions are. Walking the factory floor, you might see a few people running the machines rather than masses of labourers doing rote work. Facilities have sports centres, libraries, and nurseries. Auditors pop by unannounced. There is still work to do, but it's not like what one might assume.

Big, evil companies must just lay off thousands of factory workers to put machines in their place, right? Wrong. China is making moves to go from being considered the world's factory to its premier service provider. Companies like Walmart are capitalising on these moves and taking this once-in-a-lifetime opportunity not to replace workers with machines but to upskill workers for future employment opportunities. To get a sense of the scale of this, let's look at one of Walmart's most impactful China initiatives—the Women in Factories Program. As a consultant with BSR, I had the opportunity to work firsthand on this project.

The program's premise was simple: go into Walmart's supplier factories across China and upskill workers. Rather than approaching this like on-the-job training, we would educate workers on subjects where they may otherwise have knowledge gaps. These subjects would go beyond business and management skills to include communications, family planning, personal finance, health, and wellness. Over the eight or nine years of compulsory education in China, there is little time to teach things like personal hygiene. In a country where abortion is still considered a legitimate form of birth control, teachers don't help students understand healthier alternatives. With children left behind in villages while parents go to the city to work, learning how to bridge both the physical and mental divide is critical for development. These social skills are just as crucial to lifelong success as learning to run a machine or compose an email.

But what does any of this have to do with business? Why would Walmart pay a significant amount of money to have factory workers go through this program instead of simply teaching them how to be quicker on the line? As with all successful sustainability initiatives, the Walmart program married the feel-good altruism of helping workers with an economic imperative that made perfect business sense.

Take the abortion topic, for example. If I were to approach a middle manager at one of these factories and say, 'Hey … let's get your workers to stop having so many abortions,' they'd laugh at me and kick me out. Instead, I'd tie the pitch into how this impacts the bottom line. When a large portion of your workforce is out because they're either having the procedure or recovering from it, the factory has high absenteeism. This, in turn, slows production since workers are missing. When they return to the factory, they might suffer from after-effects, like depression, that adversely impact their work.

The results are amazing when you turn all of this on its head. Educating female workers on proper family planning means they're getting fewer abortions, showing up to work more, and are more productive. Not only that, but they're also happier, healthier, and more appreciative of the factory for investing in them. This leads to lower turnover. All told, this means more money for the factory and Walmart. Approaching the conversation this way demonstrates a true win-win scenario. How much of a win-win?

Over three years, nearly 90,000 women across fifty major Chinese factories went through the program. That's 90,000 lives changed directly and countless hundreds of thousands more indirectly. Along with acquiring the hard skills necessary for on-the-job success, participants became more self-aware, communicative, and confident. More than 70 percent of employees said the program helped them adapt better to, and solve problems in, their personal lives and at work. Eighty percent of factory trainers said their self-confidence and communication skills improved. Ninety percent showed improved communication skills.[11] Such scale and impact are something most governments, or even the best-funded NGOs, would have a hard time accomplishing. Again, this demonstrates the power of the private sector when used to its full potential.

Walmart is only one example of private sector companies doing social good at scale. Similar programs are focusing on the betterment of factory workers across much of the developing world. For over a decade, The Gap's P.A.C.E program, The HER Project—sponsored by corporations like Disney, HP, and Levi's, and Plan W from Diageo— has made a difference for millions of workers. The crazy part is few outside a very inner circle would even know such programs exist. It goes back to the idea companies are not only reticent about talking up the great things they're doing, but sustainability pros are just as lousy at marketing these efforts.

Now consider this new information and how it might impact where you shop. Would you be more inclined to choose a company running one of these programs versus its competitor? It's not as if they're asking you to pay more at the register. The programs are in place. You've been paying for them already. Weirdly, there's a bit of reverse greenwashing going on here. As companies doing good work get vilified, incorrectly in some respects by activists and unknowing consumers, our dollars might flow to businesses not having nearly as significant an impact. Until all companies compete on an equal sustainability footing, education will continue to be essential in furthering some of these worthwhile social endeavours.

Governance

If social sustainability is the sexy, fashionable side of my profession, then governance is its geeky, nerdy cousin. This is the hard numbers, spreadsheets, and reams of data part of sustainability. It's the factory-visiting, question-asking auditor. It's the pencil pusher staring over your shoulder to make sure you've dotted that I and crossed that T. Governance is also the part of sustainability where I spend most of my time. I guess that tells you a lot about how my mind works!

But contrary to what people might tell you, governance is more than just the ticking-box exercises private sector companies must undergo. That's because governance is the aspect of sustainability shoring things up against the onslaught of greenwashing. Better governance, oversight, and transparency make it increasingly difficult for companies to lie and fabricate information. Done correctly, governance makes greenwashing impossible.

If we return to our evolutionary imagery, governance is the naked man standing up straight and walking purposefully. It's typically the very last stage of the sustainability journey. Yes, there is always something more companies can do to be more sustainable. But when they have robust governance mechanisms in place, nine times out of ten, everything else is in good working order, too. These mechanisms could include, for example: linking pay with sustainability performance; having a diverse board; appointing a chief sustainability officer or similar executive role; doing multi-year reporting on ESG performance; conducting materiality analyses that help the company focus on the most critical sustainability areas to address; or having solid stakeholder-engagement practices.

Given this is the very top end of the spectrum, though, examples are few and far between. Only recently have major companies started to play around with tying remuneration to sustainability KPIs. The most notable and aggressive of these are Alcoa (who was also one of the first major companies to do so), Apple, Pepsi, and Unilever.[12]

Often hailed as an exemplar for corporate sustainability, Unilever's long-standing CEO Paul Polman received a hefty US$772,000 bonus in 2014. This bonus was entirely tied to the company's industry-leading Sustainable Living Plan and was on top of an already eye-wateringly high US$1.4 million payout. Apple raises and lowers executive payouts by 10 percent based on ESG performance. Going even further, Intel ties bonus schemes across all levels of the organisation to ESG metrics.

Leadership consultancy Semler Brossy began tracking progress of the Fortune 200 on linking executive compensation to ESG back in 2019. Since then, they've expanded their tracker to cover the S&P 500. These are the 500 largest publicly traded companies in the United States. Their most recent report for 2021 found 57 percent of S&P 500 companies '... disclose some form of ESG to determine incentive compensation outcomes.'[13] One would assume most of these link to environmental metrics. Interestingly, though, 28 percent of companies had diversity and inclusion as their top metric. For those wondering, this falls squarely in the social part of ESG. Unfortunately, there isn't another ESG metric until company culture ranked seventh at 13 percent. Then it's a pretty precipitous drop to environmental metrics, most of which hover around 2–3 percent.

While a move in the right direction, having 'some form' of ESG doesn't give me the strongest confidence. It's the kind of flim-flam that screams greenwashing. I'd love to see even more concrete incorporation of ESG metrics, with a concerted long-term strategy, from businesses in the future. It's not like this is a big ask, either. According to a recent PwC survey, 94 percent of board directors want the inclusion of non-financial goals (like sustainability) in an executive's compensation plan.[14] With increasing investor attention on greening portfolios, they also want executives to be accountable to more than just profit. The 2021 *Global Benchmark Policy Survey* discovered that '... 86% of investors (and 73% of non-investors) think non-financial ESG metrics are an appropriate measure to incentivize executives.'[15]

Let's assume a company doesn't want to compensate its executives based on sustainability performance. What then? Since many boards are still primarily pale, male, and stale, they could put a focus on diversity. This diversity and inclusion metric is already permeating much of the S&P 500. All a company would have to do is float this up to the executive and board levels.

If we look at board representation, there are several ways to diversify. A company could be more gender diverse, ethnically diverse, or age diverse. It could also bring in non-executive directors with specialisations, like sustainability, to add diversity of thought. A pre-Covid survey of financial institution board diversity in the US named Ohio-based Fifth Third Bank the industry's most diverse board. Its composition is as follows:

> … one-third female and one-quarter who are considered ethnically diverse, with one younger director. US Bancorp, which also placed second in Bank Director's 2018 RankingBanking study, has a board composition that is 29 percent female and 21 percent ethnically diverse. An impressive 29 percent are younger directors.[16]

Beyond finance, there are other private sector leaders in board diversity. General Motors has a board comprised of 58 percent women, while Proctor & Gamble is a close second with 50 percent female representation.[17] In Europe, DSM, Kering, and Diageo all have boards with at least 50 percent women.[18] Racial diversity is also an increasingly important focus area. While not where most probably want it to be, companies like Starbucks, Accenture, and Target have boards of around 50 percent non-white directors.[19] Getting to this point has definitely taken a lot of effort, planning, and persistence. Starbucks, for example, began their board diversification journey back in 2017. Since then, they've had a board chair publicly championing the pursuit of a racially diverse board. Accenture's CEO, Julie Sweet, is also vocal about diversity's benefits at the executive and board levels.

In a 2020 LinkedIn post, Sweet doubles down on making Accenture '... more inclusive, diverse and representative of the communities where our people are living and working.'[20]

Still, others lag well behind their peers. The Australian Institute of Company Directors notes that while all ASX 200 boards have at least one female member, nearly 70 percent of board seats still go to men.[21] With only about 40 percent of new board appointments being women, this doesn't look likely to change. Staying in Australia, the *Financial Review* revealed 90 percent of all newly launched companies between 2019 and 2021 had all-male boards.[22] Looking at the most recent Gender Diversity Index by the organisation European Women on Boards,[23] we find EU countries average 35 percent female board membership, with Norway and France topping out at mid-40 percent. Greece and Luxembourg sit at the other end of the list, with 24 percent and 18 percent female representation, respectively.

Turning to the world's biggest players, only 31 percent of Fortune 500 board members are female, and just 7 percent non-white women.[24] This is even more shocking when you discover 2021 was a record year for female board appointments. Female, minority-owned business, Mogul, crunched the numbers and also found that '... 16 companies within the Fortune 500 have no ethnic minorities on their boards; three have no females or minorities; and none of the boards have Native Americans on them, male or female.'[25] Where this is the worst is in the tech sector.[26] Tesla, Intel, Cisco, and Amazon have only about 20 percent female leadership. Shopify, Zillow, and eBay have less than 7 percent racial diversity in their leadership roles. Some household tech names who don't even bother to report on gender or racial diversity in leadership (likely because there's not much to report) include NVIDIA, Oracle, Adobe, IBM, SpaceX, Qualcomm, and Bloomberg. So the next time any of these companies want to talk diversity, know they're trying to misdirect you away from the problems in their own boardrooms.

A relatively new kid on the block, the Chief Sustainability Officer role is also an excellent way to ensure strong ESG governance across an organisation. DuPont appointed the first CSO at a public company, Linda Fisher, in 2011. Growing in prominence, the number of CSO appointments tripled in 2021.[27] Interestingly, and contrary to the actual title, only about 28 percent of these are at an executive level. Most are senior VPs or higher-level managers. The key to understanding their remit is to juxtapose the role against all the other sustainability endeavours a company is (or isn't) doing.

The role has also grown in notoriety, for better and for worse. *Sustainability Magazine* recently published its Top 100 Leaders in Sustainability[28] edition, a list comprised almost entirely of CSOs from the world's biggest companies. The list did include some forward-thinking CSOs, like Kate Brandt from Google, Rebecca Marmot of Unilever, and Werner Baumann of Bayer. Unfortunately, it's a hard pill to swallow when most of the list were CSOs from some of the most infamous greenwashers. There's Bea Perez from the world's biggest plastic polluter, Coca-Cola. You've got CSOs from oil and gas companies like Petronas. There were even a few CSOs from defence contractors, including Mike Witt of Northrop Grumman, one of the world's largest weapons manufacturers. I'm sure his working hard on the '... design and execution of enterprise-wide business activities, including carbon management, and improving resource and material management' will be a comfort for those watching their villages get hit by a missile.

Like our evolutionary imagery, corporate sustainability has undoubtedly shifted and changed over the past twenty years. It's outdated to associate sustainability simply with philanthropy, going green, or being a good corporate citizen. Today, sustainability is an umbrella term for anything building a more positive future. In the corporate context and through the lens of ESG, it covers pretty much everything. Sustainability is no longer a part of how companies do business. Now, sustainability is the business itself.

Yet, the highly matrixed nature of sustainability inherently means there are companies at different stages of development. Some industries are seasoned and mature, while others are only starting out. Even within sectors, there are vast differences in approach, focal areas, and understanding between companies. Geography makes this even more complex. Sustainability work in the US and EU is significantly different from that done in developing markets like China, Indonesia, and India.

What does all that have to do with greenwashing? As companies continue to evolve and become more sustainable organisations, the available space for greenwashing will increasingly shrink. As we've seen throughout this section, innovation will drive competition. But unlike the traditional capitalist dog-eat-dog scenario, this innovation will place the private sector in a virtuous cycle. They will innovate to be better. By marrying profit with planet, they will see how lucrative, differentiated, and competitive they can be. They'll also reach a critical mass where companies that don't play ball will no longer have a role.

'Are we there yet?' asks the petulant five-year-old in you from the backseat.

Not quite … but we're not far off.

Private sector companies on the vanguard have already laid solid foundations followers can build off. Whether that's completing the research necessary to green the auto sector, developing happy, healthy, productive employees, or building diverse boards to drive innovation forward, the solutions exist. There is absolutely no reason to reinvent the wheel. The work now is for companies that may be late to the party to jump in and get started. Those companies that can change are changing. For any hesitating or resistant, there's not much runway left.

In the infamous words of Richard Nixon, now's the time to shit or get off the pot.

Chapter 8

Fast Fashion

Aaditri slowly opens her eyes. The stifling heat and humidity of the past few nights have made sleeping difficult. New to the city, she's still not used to all the strange noises. The countryside may not be as developed, but at least there's plenty of peace and quiet. There she could get a good night's sleep. Maybe once she's saved up a bit of money, she could go back and visit her family. Even though nearly 21 million people surround her, Dhaka can still get lonely.

Scratching at half a dozen mosquito bites on her arm, Aaditri sits up and greets the day. It's still dark outside. She has to get up early to travel the 15 miles from her small, shared apartment in Gulshan to the garment factory hub of Savar. This new suburb on the outskirts of Dhaka only sprang up in the last few decades as land prices in the city centre skyrocketed. Reclaimed marshland became dusty industrial parks to serve the never-ending flow of orders from western retailers. There are 4 million Bangladeshi garment factory workers, 58 percent of whom are women.[1] Textiles make up 80 percent of the country's total exports. Even with a pandemic slowdown, that number stands at over $35 billion, second only to China.[2]

Aaditri creeps outside quietly to avoid waking her flatmates. She goes down her building's concrete stairs and finds ankle-deep water; the remnants of last night's monsoon. A dramatic influx of migrants, coupled with climate change, is wreaking havoc on Dhaka's inefficient public systems. One of these weak systems is the aqueduct network, which helps to alleviate the annual flooding of the city's rivers. In the rainy season, it's not uncommon for one-fifth of the country to flood. Climate change is dramatically changing weather patterns at the exact time the country can least handle them. Over the past decade, nearly one million people in Bangladesh have been displaced annually due to climate-related catastrophes. Experts estimate this will rise to 13 million annually by 2050. Women and girls, like Aaditri, are particularly vulnerable to these and other changes.[3]

The sun is barely coming up, but the bus to Savar is already packed. Aaditri jams in as best she can, clutching her bag tightly against her chest and keeping to herself. She puts on her headphones to drown out the cacophony filling the bus. This is as quiet as things will probably be today, so she takes a minute to enjoy it. The bus makes its final turn into Savar, stopping on Mazidpur Road. Here, most straphangers get up and off. Some will walk north towards the Bazar. Others, like Aaditri, will cross the highway and walk south.

They'll pass rows of dilapidated buildings; eight, nine, and ten stories tall. But about halfway between the bus stop and Aaditri's factory is a large empty plot. The space is overgrown with grass and weeds as the once-fertile marshland tries to reclaim its glory. On the left is a high wall of white concrete covered in advertisements. To the right is a garish building painted with blue, pink, and mustard-yellow stripes. Aaditri remembers the first time she passed this spot. It's one of the few open spaces in the city, but she noticed nobody ever stepped foot on it. In asking around the factory, she quickly realised this was once the site of Rana Plaza.

Rana Plaza was an eight-story commercial building right in the heart of the Savar district. Outside, it was no different from other buildings housing banks, restaurants, and garment factories. Like its neighbours, Rana Plaza sprang up in the development frenzy around 2004. Hundreds of similar buildings went up around the capital, all with lax regulation, shoddy materials, and poor planning. Several, like the Spectrum Building, had already collapsed. While the Government briefly toyed with the idea of cracking down on these unsafe practices, money streaming in put safety on the back burner. This became even more transparent during the Global Financial Crisis as western brands put increasing pressure on factories for cheaper, faster production.

Executives in Paris, New York, and London kept an eye on the bottom line. Some sustainability teams started shoring up pay and labour issues throughout their supply chains. Few, if any, were paying attention to the factories themselves. By 2013, this was a ticking time bomb.

Then the inevitable, but entirely avoidable, happened.

Problems started on the afternoon of April 23, 2013. It was then workers noticed major cracks appearing on the building's support pillars. In one of the few wise decisions taken, local government officials immediately shut down the retail operations occupying the lower floors and the five garment factories on the second through seventh floors. With everyone sent home, surveyors could assess further.

After another inspection overnight, the building got the all-clear. It was as if the structural cracks magically disappeared on their own.

To understand the *why* behind this blatantly illogical decision, one has to grasp the pressure put on these factories by brands. A common practice at the time was for buyers to deduct a percentage of pay every week a shipment arrived late. With razor-thin margins, this could spell disaster for a factory and its workers. Any time off work made this threat a more palpable reality.

The interconnected yet highly complex nature of garment production also meant factory managers needed everyone at all times. One year after the tragedy, a write-up in *The Guardian* explained this fact well. In one factory, Phantom Apparels, a garment worker named Mahmuda had to stitch seams and pockets: 'It is one of the tougher of the 60 or 70 separate operations needed to make a pair of trousers. She is supposed to stitch 120 pairs an hour, 10 hours a day, six days a week, 50 weeks of the year. That's 360,000 annually.'[4] She earns US$78 a month for her back-breaking work. If she's out sick or injured, she earns nothing. Her absence also negatively impacts production schedules.

On the morning of April 24, less than twenty-four hours after cracks started to appear, 3122 factory workers were back at their stations inside Rana Plaza. Some managers used a carrot to entice cautious workers back into the building. Others, like Ether Tex, threatened to withhold a month's pay for those who didn't return. Work started at 8am. By 8.30, the building's power had gone out. This wasn't unusual, but it worsened the already suffocating conditions inside the factories. As the rooftop generators fired up, they sent significant vibrations down throughout the building.

Within minutes, one of the building's structural corner support pillars gave way. The explosion was loud enough that most would have known what had happened. By then, though, it was too late. Each floor began to list and dip like a sinking ship. Workers on one side of the building would have been thrown up in the air, only to slide down towards the other. All the while, heavy machinery came crashing down around and onto them, breaking bones and knocking many workers unconscious. The third floor, where Mahmuda worked, gave way instantaneously. This led to the eighth-floor caving onto the seventh, the seventh onto the sixth, until only the bottom floor remained. Much like the World Trade Center in New York, Rana Plaza was a complete structural failure.

The whole collapse happened in less than ninety seconds.

International teams led by the United Nations offered search and rescue assistance. These offers, though, were flatly rejected by the Bangladeshi Government in an attempt to save face. Ultimately, efforts fell on an ill-equipped team primarily of volunteers in sandals and without protective equipment. Pressure from family members kept the rescue work going when the Government tried to end things quickly. These missteps likely led to the loss of many more lives. After nineteen days of search and rescue, teams recovered 2500 survivors. The final death toll stands at 1134 people. This makes Rana Plaza history's deadliest garment-factory disaster.

No one event, person, or company led to the Rana Plaza disaster. To be honest, we all played a hand in the tragedy. That's because we've become accustomed to having what we want when we want it, especially with our clothing. You can pick up an entire suit at the Tokyo Airport waiting for your flight. Or pop down to the shops for a nearly designer dress on your lunch break. It now takes remarkable willpower to window shop at the mall. With low prices and new designs constantly churned out, most people would instead step inside the stores and spend, spend, spend.

The numbers back this up. An average person in the developed world today will purchase over sixty items of new clothing every year, according to Rent the Runway.[5] This adds to their wardrobes, which usually consist of no less than a further sixty to seventy items. What's worse is that most will only wear each piece a few times. In a pre-pandemic article from *The Wall Street Journal*, '… on average—average—each piece will be worn seven times before getting tossed … [i]n China, it's just three times.' We're buying more today than ever before while chucking things out before they even have time to wear in properly.

All you have to do is think about your own closet. For the brave, I'd challenge you to put down this book and go count the number of items you have sitting in there right now. It's okay. I'll wait.

How'd you go?

If you counted around 120 items, then you're pretty average. And don't worry about all those tattered undies, misshapen bras, and mismatched socks. We won't count those towards the total. While you were counting, did you come across a few things you totally forgot you had? How much stuff do you actually wear in a given month? I guess you have a few favourite pieces you wear repeatedly. That, too, is pretty normal. A study of American women found the average person only wears 10 percent of what they own.[6] The other 90 percent of your wardrobe is just sitting there collecting dust.

But it hasn't always been like this. In fact, the concept of fast fashion as we know it today is only a few decades old. You don't have to go too far back in history to see made-to-measure, seasonal clothing as the norm. Many of us still come from that era. Many of our mothers and grandmothers would have made their own prom dresses hunched over a Singer sewing machine. I'm just a few short years over forty, and my mom cut cloth diapers by hand for me when I was a baby. Slow fashion shouldn't be so foreign to us.

That's because for a wide swath of human history, making clothing was pretty much the same. Up until the 1800s, this was a laborious process. The cloth didn't come on bolts, ready-made and in a variety of colours. One had to either make or source the material themselves. That could require undertaking the disgusting process of tanning leather, which uses a lot of urine. Or, you might have to wait until the end of winter to sheer your sheep (then spend months cleaning, carding, spinning, and weaving the wool). Maybe, you'd need to spend time praying the latest cotton crop is good enough to make a nice pair of knickers (while turning a blind eye to who harvested it).

Oh, you wanted something a little brighter than just dull yellow? Since natural dyes had to come from organic sources, most of us were limited to what was nearby and what we could afford. That could mean crushing hundreds of brightly coloured bugs, or extracting indigo from its namesake plant, to find the perfect colour for that top or dress. As many of us would have learned in history class, the most prized colour was purple. This colour (Tyrian Purple, to be more precise) was associated with royalty not because it was the prettiest but because it was the most expensive to produce.

What you may not have learned is how backbreaking the production process was. You can only create Tyrian Purple from the Murex, a predatory sea snail native to the Mediterranean and Atlantic. These pretty boring-looking creatures secrete mucus that can knock out their prey or as a defence when attacked. When this mucus hits oxygen, it turns purple. To get there, though, you have to collect enough snails and then either: a) continuously poke and agitate them, so they secrete that prized mucus; or b) destructively crush up the creatures. In an early example of humans saying 'screw it' to a renewable resource, we chose the latter option. Once destroyed, you had to have someone with enough background in science to take the mucus through the requisite biochemical reactions and finally to the dyeing mill.

Do this with close to 12,000 snails, and you'll have 1.4 grams of Tyrian Purple, enough to dye the trim of a single garment.[7] With all this in mind, it's no wonder our ancestors weren't stocking up on sixty, seventy, or eighty new items in a year.

Things started to change during the Industrial Revolution. It was in 1755 when Charles Weisenthal received a British patent for a 'needle that is designed for a machine'. Over the next forty years, many people iterated the invention. Eventually, French tailor, Barthelemy Thimonnier, created the world's first functioning sewing machine in 1830.[8] Originally designed to produce uniforms for the French army, his machine became something of a sore spot for fellow tailors. In a spate of anger and fear the humble sewing machine would put them all out of a job, they burned down Thimonnier's factory with him inside. He survived, as did his machine, starting our long march towards fast fashion.

What the sewing machine did was create opportunities for clothing production at scale. Instead of a single woman (yes, it was always a woman) having to stitch something by hand, entire textile factories could now make garments quicker, cheaper, and easier than ever before. Even the local dressmaker could ramp up their production for the middle classes. With the sewing machine, they began to hire workroom employees. They also started to outsource tasks to those on low wages working from home. These 'sweaters' as they came to be known, would eventually lend their moniker to the infamous sweatshops associated with the industry today. While still early days, this was the beginning of a mass-production mindset towards ready-made clothing.

But technological innovations weren't the only thing spurring greater textile production. The North Atlantic slave trade, primarily fuelled by cotton, provided a sinister way to reduce production costs further. Before the American Civil War in 1861, cotton was Britain's largest industry. Over 80 percent of that cotton came from the US, creating the infamous triangular route we associate with the slave trade. More textiles created a need for more cotton, which in turn created a demand in the mind of traders for more slaves. The vicious cycle continued throughout this period. Even after the Civil War and abolition of slavery, continuing colonialism exacerbated a structural dependency between coloniser and colonised. This pattern of development between the industrialised and agricultural worlds is still very much with us today.

Developments in the textile industry continued apace throughout the Industrial Revolution and up through the mid-1900s. When I talk about developments, these relate to how things were produced, not the people who produced them. Although the Atlantic slave trade was well and truly over, slavery itself had just put on a different outfit. A little over one hundred years ago, immigrants from all over Europe crowded ships bound for New York Harbor. All were fleeing different things—war, famine, persecution—but the commonality among them all was hope for a brighter future. The 'new world', with streets lined of gold, would offer opportunities beyond their wildest dreams.

While in transit, these immigrants heard stories of the great families of American history undergoing the same dangerous journey: the Carnegies, Rockefellers, and Morgans. These families were also immigrants who went from rags to riches in the most American way possible. Although these people, and others like them, received most of the headlines, they were far from the norm in the US at the time. In 1910, a decade after the rapid economic growth of the Gilded Age, the average American worker could expect to bring home about $750 per year. Adjusted for inflation, this would equal a little over $20,000 today.[9] Remember, though, most immigrants weren't making nearly this much. Those immigrants, if they could make it past inspection on Ellis Island, would probably end up in crowded, rat-infested tenements in the Lower East Side. They were likely to spend their days in a hot, cramped, and dangerous warehouse, operating machines that could quickly turn into torture devices for misplaced limbs. If they injured themselves, it was likely to be game over. They could attempt to rely on the charity of others, yet given how stretched most budgets were during this era, it was probably the end of their American dream.

On a warm spring afternoon in 1911, the dreams of 146 American immigrants would come to an abrupt end. The Asch Building, a block or two east of Washington Square Park in New York's Greenwich Village, would become the site of one of the worst industrial disasters in American history. On the building's top three floors, the 500 employees of the Triangle Shirtwaist Factory were finishing up work for the day. Many of the workers were young women, recently immigrated from Italy and Eastern Europe. Although we would deem it a sweatshop today, most considered this immigrant-owned business the height of efficiency for the time. Many were happy to have landed such a job when they knew others across the city weren't nearly so lucky.

It was a Saturday, so the short seven-hour day ended at 5pm. Some would have glanced at their wristwatches reading 4.40pm. and begun to clean up their stations. The only thing left was to line up for inspection; a pesky task to ensure nobody was stealing anything. That's why they always kept the doors locked. The owners didn't want anyone running off with their goods.

Nobody saw when or how it started, but soon smoke began filling the rooms. Looking around, the women on the eighth floor would have noticed flames coming from a large wooden bin under one of the worker's tables. Maybe it was a match or a cigarette butt. Perhaps an engine from one of the sewing machines was the culprit. Or, more nefariously, did one of the owners do this to collect on the insurance? No matter what ignited it, the hundreds of pounds of scrap cloth accumulated for months under those tables were aflame within five minutes. Fire nipped at any flammable surface, which in this factory was everywhere one looked.

Some workers on the eighth and tenth floors could use the elevator or take the stairs to the roof. Those on the ninth floor, though, were not so fortunate. There was no intercom or telephone system on that floor. Even though it started a floor below, the workers on floor nine only found out about the fire as it began to engulf them. By then, anyone with keys to the locked doors had already fled. The stairway down to Greene Street was on fire. Taking their chances, dozens of workers clambered out onto the fire escape. The slap-dash structure buckled under the weight. As onlookers filled the streets below, they would have watched in horror as the workers' bodies and twisted wrought iron began raining down. One reporter on scene remarked he had, '… learned a new sound that day … the thud of a speeding living body on a stone sidewalk.'[10]

A New York State Assemblyman, Louis Waldman, was also watching from a nearby library. His firsthand account captures the sheer terror of the event.

> A few blocks away, the Asch Building at the corner of Washington Place and Greene Street was ablaze. When we arrived at the scene, the police had thrown up a cordon around the area and the firemen were helplessly fighting the blaze. The eighth, ninth, and tenth stories of the building were now an enormous roaring cornice of flames.

Word had spread through the East Side, by some magic of terror, that the plant of the Triangle Waist Company was on fire and that several hundred workers were trapped. Horrified and helpless, the crowds—I among them—looked up at the burning building, saw girl after girl appear at the reddened windows, pause for a terrified moment, and then leap to the pavement below, to land as mangled, bloody pulp. This went on for what seemed a ghastly eternity. Occasionally a girl who had hesitated too long was licked by pursuing flames and, screaming with clothing and hair ablaze, plunged like a living torch to the street. Life nets held by the firemen were torn by the impact of the falling bodies.[11]

In less than an hour, the lives of 123 women and twenty-three men came to an end. For ninety years, the Triangle Shirtwaist Factory tragedy held the unfortunate title of deadliest workplace incident in the United States. It is only eclipsed by the September 11 attacks.

· But from this tragedy came much-needed reform. Labour rights, including the right to unionise, and the beginnings of occupational health and safety reform, are all legacies of the Triangle Shirtwaist Factory. With the Industrial Revolution coming to a close, workers were now demanding their rights as human beings instead of just a means of production. As with Rana Plaza over a century later, these horrific incidents were wake-up calls—albeit far too late—for change.

With the end of World War Two, thinking towards fashion began to change again. The industry still produced clothes in seasons as it always had. Yet frugality, psychologically embedded into many following a decade of war rations, meant anything one purchased had to be cost-effective. Mass-produced clothing began to overshadow made-to-measure for most middle class consumers. Fashion historians clarify this period, though, as still one with a distinction between high-end and high-street.

That all started to change in the swinging '60s as younger consumers began expressing their individuality by rejecting fashion norms. The biggest change? Looking good didn't have to come at a cost. While their frugal parents were okay with cookie-cutter outfits, this generation embraced cheaply made, but hip, clothing. Companies tried to keep up, but now had to create high-end, trendy looks while keeping costs as low as possible. Some even started to look further afield at less expensive locations in South-East Asia and Latin America to set up shop. If all this sounds familiar, that's because it's the fast fashion model we've become accustomed to today.

Right now, you're probably asking yourself who the first fast fashion company was. It makes sense to try and pin the blame on a single bad apple. If we're going simply by timeline, H&M is the oldest of today's fast fashion giants. Founded in 1946 by Erling Persson, his inspiration for Hennes & Mauritz came from large retail establishments in the United States. Initially, the stores only sold women's clothing. Over the next thirty-plus years, the company kept a predominantly Scandinavian footprint. Their most significant expansion during this period came with the 1968 acquisition of a men's hunting apparel retailer. Other than that, you'd be hard-pressed to find anyone in Europe or the US who knew about H&M. While H&M would grow to become the world's largest fast fashion retailer, their humble beginnings are pretty blameless.

In my opinion, the more important question is, *who started the fast fashion phenomenon?*

The biggest contender for that title would have to be Spanish retailer Zara.

This has nothing to do with when the company was founded. Zara didn't open retail locations until almost thirty years after H&M. Zara takes the cake for being the first true fast fashion company because of its business model. From the very outset, founder Amancio Ortega aimed to create high-end fashion at low-end prices. His differentiator, though, would be speed. This need for speed eventually started a race to the bottom for the planet and, as we'll see shortly, the people who have to make all this stuff.

I hoped to accomplish two things in providing this whirlwind history lesson of the fashion industry. First, I wanted to demonstrate that as much as we think fast fashion has always been a part of life, this couldn't be further from the truth. Much like humanity itself, fast fashion is just a blip on the historical radar. As I said, we're less than one generation removed from seasonal, made-at-home clothing. This isn't to say I want us all to go back to our Singer Sewing Machines and start churning out clothes. Instead, there's no reason we can't have a bit more balance than what we see today.

My second goal in breaking down history is to show just how removed from reality sustainability claims in the fast fashion industry actually are. No amount of organic cotton can offset a building collapsing on top of your workers. How many recycling boxes make up for the millions of gallons of water used to make a new pair of jeans? If you're scratching your head wondering when the greenwashing bit will start, now's the time.

Moving away from the history of the fast fashion industry, I want to now turn to the three major mechanisms in its greenwashing machine. The first piece of this puzzle relates to the psychology of stuff. Basically, how they get us to buy so much when we never felt the need before. All those shiny new products have to come from somewhere. This leads to the second mechanism: resource use. But cotton doesn't sew itself. That's why we also must consider the labour involved in making the clothes we desperately desire.

These three mechanisms combine to create an entirely unsustainable industry, no matter what flashy marketing speak you might come across. Yet some fast-moving consumer goods companies have gone on a different path. Instead of take, take, take, they are looking to balance the picture. We'll also look at these more positive cases as examples of where fast fashion could evolve.

The Psychology of Stuff

Now, don't get me wrong. It's not as if I'm better than all of you. I'm certainly not immune to the lure of fast fashion. In fact, I'd be the first to tell you it's been a bit of an addiction for a very long time. The thing that always gets me? *Sweaters.* Until pretty recently, I had boxes and boxes of sweaters. At one point or another, I just had to have them. God forbid I passed by a store with a winter stock-take sale. That's when I would make some terrible financial decisions and countless fashion *faux pas.* Even if I grew out of something, I'd keep it as my 'inspiration item'. You know what I'm talking about, right? The piece of clothing that's going to get you back to that summer bikini body or inspire you to shred those abs. Spoiler alert: they never did. After years and years, the sweaters just kept piling up. They'd follow me around from house to house, country to country, as I travelled the world.

Where did this all start? Well, early on—and we're talking almost twenty years ago—I'd love to head down to Soho for a little shopping spree at Uniqlo or Top Shop. As a young twenty-something in New York City, having the latest and greatest was paramount. Seeing how these shops were just hitting the city, they were novel, cool, and dare I say, fashion-forward. Nobody thought twice about the amount of stuff stocked on the shelves. Friends weren't going to turn their noses up when you told them about your new H&M outfit. Even professionally, fast-fashion retailers were filling a space somewhere between Hugo Boss (good luck affording a suit as a student without a trust fund) and Burlington Coat Factory (more than great coats, sure, if you're a size XXXL).

Soon, though, the shine started to wear off. Friend and comedian Ophira Eisenberg hit the nail on the head when she joked about relationships. Some are great and destined to go the distance. Others are like that really cute top from H&M. It might look amazing now, but you just know it's going to fall apart in six months.

They say comedy holds a mirror up to reality, and boy did Ophira's joke hit home. I laughed until I realised how true the statement was. I'm sure I looked down to double check a button hadn't fallen off.

A real come-to-Jesus moment for me happened during BSR's annual sustainability conference. While a consultant with the organization, I attended a few of these conferences. I always left inspired and encouraged. As one of the world's biggest gatherings of sustainability professionals, it's no surprise the calibre of activities one can access. Some of my favourite moments from these conferences included listening to the likes of ultra-marathoner Mina Guli as she described running twenty-nine marathons in thirty-eight days, over seven deserts on seven continents, to raise awareness for water conservation. There was also Pascal Finette from Singularity University, who explained the impact exponential information growth has on accessibility. Most poignant was Lord John Browne, former CEO of BP. In speaking on authenticity, he recounted his harrowing journey dealing with media harassment that eventually forced him out of the closet and out of the company he led for twelve years.

Behind the scenes, there were also plenty of opportunities to connect with colleagues. Coming from the offices in China, I especially liked interacting with people I knew only through a computer screen. During one notable icebreaking session, I was seated at a table of around ten people. As luck would have it, that table included BSR's CEO, CFO, and a couple of MDs. Although an excellent opportunity for face time with the powers-that-be, one particular icebreaker question really gave me pause. What was our sustainability guilty pleasure? While others took a few minutes to gather their thoughts on the question, I immediately knew my answer.

Going around the table, you had some of your stock-standard responses:

'Even though I know water is precious, I love to take long, hot showers.'

'I've really got to take public transportation more. I'm way too reliant on my car.'

'I don't volunteer enough, even when I have spare time.'

All of these elicited understanding nods and a bit of commiseration. Who out of us doesn't need to reduce our shower time or help out a worthwhile charity? The CEO was seated next to me, and his response always stuck. Visiting BSR offices worldwide, being on speaking circuits, and working with global clients meant he was constantly on and off planes. Although he knew how bad this was, even if it was less a guilty pleasure and more a necessity of the job, he was actively looking at ways of offsetting his environmental impact. Maybe because it was the CEO, or perhaps because people could relate, he also received the requisite nodding of heads.

Then it was my turn.

'My sustainability guilty pleasure is shopping at Uniqlo.'

Now, dear reader, perhaps my mind has made this moment something more than what it actually was. But I swear to you, there were audible gasps from some of my colleagues. Heads went from nodding up and down to shaking left and right. I could feel their eyes scanning me, trying to identify which terrible fashion brand had made my suit (joke's on them ... it was from a bespoke tailor in Shanghai!). How in the world could my shopping sprees be that much worse than spewing out diesel fuel flying over Nebraska? Surely, riding your bike two days a week to the office wasn't really offsetting joyrides the other five days. To this day, I still don't know what elicited such a response to my comment when the others just went by without a second glance.

What was clear, though, was just how far fast fashion had fallen out of favour. In less than a decade, we had gone from people frothing over the latest Mango top to being visibly appalled just by association. Mentioning such a guilty pleasure, especially among a group like those at the table, was pure taboo.

It was also around this time scientists began to study the neurological impacts of fast fashion, with many likening it to an actual addiction. Why? As you'd probably expect, that devilish little chemical dopamine pops up in the study. That's because, just like snorting a line in the toilet of the city's hottest new club, shopping makes you feel good. It gives you a high, and you can't get enough. A 2007 study by a group of American researchers explicitly explored the impacts of shopping on the brain. Participants saw various items for sale. The brain's pleasure centres activated when they got to an item they liked. This activity increased with the item's perceived desirability.

Interestingly, the researchers found it's not just purchasing items that leads to dopamine release: 'Together, these findings implicate mesolimbic dopamine projections areas in the representation of anticipated gain, but do not clarify different roles of these distinct projection areas.'[12] In plain English, it doesn't matter if you're buying something or just thinking about buying something. Even the simple act of window shopping, physical or virtual, can increase dopamine levels. Getting a great deal on something you really love opens those dopamine floodgates as well.

In economics, this is called transactional utility. The pleasure you get from buying something isn't just in the product itself but also in the bargain that might come with it. Fast fashion retailers are genius manipulators of this principle. One of the best examples I can think of comes from Uniqlo, *quelle surprise*. The company likes to partner with prominent name designers to release special edition collaborations. Some of the most famous collaborations have been with Kaws, Jil Sander, Keith Haring, Christophe Lemaire, JW Anderson, and J. Lindeberg.

Suppose I was feeling my oats and popped into a Christophe Lemaire men's store in Paris. I've got my eye on a simple white cotton shirt with a lovely little button embellishment on the collar. The sleeves are just a little longer than normal, hitting right under my elbows, and oh so chic. Turning over the price tag, a whopping

US$430 is staring back at me. How about a JW Anderson boutique in Gangnam? A screen print white t-shirt there runs over US$200. A pair of J. Lindeberg men's tights from their Oslo flagship retail at US$140.

Enter the Uniqlo collaboration. Now I can go into a store and find a nice piece by one of these high-end designers. We're not talking about designer imposters, either—their names are actually on the tags! What I'm looking at might not be exactly what they have in their expensive boutiques, but it's pretty damned close. Even better, I'm not spending an entire pay cheque on a single piece of clothing. That t-shirt or dress will be well under a hundred bucks.

This is transactional utility at work.

But it's not just great design and low cost that factor into this. We can also add supposed altruism as a third part of the transactional utility equation. When these retailers push their greenwashing messaging, consumers feel they're doing good through their purchases, regardless of legitimacy. Why wouldn't I buy those jeans knowing part of the proceeds goes to some random charity in Africa? I'm getting a great deal and supporting a good cause. See, fast fashion isn't all that bad! Yet throughout this book, you'll see how often these messages are full of half-truths and obfuscated details.

And what of my own addiction to fast fashion today? Am I still mainlining it like I was before? As I've grown older and with much more disposable income than when I first moved to New York, I've gone through a natural detox. I've developed an appreciation for quality over quantity. That doesn't mean I'm going to the atelier for *haute couture*. Far from it. It just means I see no need to fill my closet to the brim anymore. Why buy something if it's never going to see the light of day?

Now I'm making smarter decisions. I opt to spend more on clothes I know will last longer than six months. Not only are they better manufactured, but I'm also confident their labour conditions are better. I recognise I am in an advantageous position compared to others, especially those who can't afford much to begin with and rely on inexpensive items from fast fashion retailers. But I genuinely believe that if you can, do. If you can't, don't worry.

I wouldn't say I'm fully cured—not that I really want to be. Fast fashion has its place in my life (I'm looking at you, undies), but to a lesser degree than it once did.

Resource Use

I'm not sure when I first noticed them. Most seemed to spring up overnight, like someone had gotten them wet or fed them after midnight. Often, they're hiding behind a pole, blending in and easy to miss. Other times they sit nearer the front in a place of prominence. Some are banded together with a bunch of mismatched packing tape. Some are crisp and brand new. Most are banged up, the result of dozens of people running into them. All, though, are seldom used.

Oh, sorry. I'm talking, of course, about those 'recycle your clothes here' bins in most fast fashion stores. You've probably noticed them, too. Formally known as take-back schemes, the concept is pretty simple. You bring in your gently-used clothing. The store then takes these clothes and magically recycles them into something new. You might get some type of voucher to use in-store for your efforts. A win for you, the business, and the environment.

At least, that's what these companies would like you to believe.

While great on paper, the real question is whether these schemes offset the resources fast fashion uses. Jeans are an oft-cited example of how much natural capital is required to make clothing. That includes the organic materials in a piece of clothing and all the resources needed to grow that material. The UN estimates it takes about 10,000 litres of water to make the kilo of cotton used in each pair of jeans. Perversely, cotton is often grown in water-starved areas.[13]

In one of the industry's first lifecycle assessments, Levi's examined the impact its famous 501 jeans had on water and energy. They found each pair will …

> … produce the equivalent of 33.4kg of carbon dioxide equivalent across its entire lifespan—about the same as driving 69 miles in the average US car. Just over a third of those emissions come from the fibre and fabric production, while another 8% is from cutting, sewing and finishing the jeans. Packaging, transport and retail accounts for 16% of the emissions while the remaining 40% is from consumer use—mainly from washing the jeans—and disposal in landfill.[14]

It goes way beyond water, though. The fast fashion industry harms other parts of the environment, too. Textile production releases over 1 billion tons of greenhouse gases each year, making up 10 percent of all global emissions.[15] Compare this to the 17 percent of global emissions from agriculture, universally derided for contributing to pollution, and it's clear fast fashion isn't far behind.[16] The fashion industry also consumes more energy than shipping and aviation combined.[17] Shipping *and* aviation—two industries synonymous with pollution and environmental degradation. Chew on that for a minute.

Then, there's the issue of clothing disposal. Can one bin, even if placed in every single store around the world, make a true impact? I'm going to say no. In Australia, for example, 6000 kilograms of clothing and textiles hit landfills every ten minutes.[18] In the United States, the world's most wasteful clothing market, the volume of clothing thrown out each year has doubled since 2000. Now, it stands at around 17 million tons of textile waste each and every year.[19] At a global level, we're throwing out nearly 100 million tons of textiles annually. By 2030, that number will rise to 134 million tons a year.[20]

The last time I saw one of those bins, it definitely wasn't overflowing with thousands of kilos of clothes. If anything, the cardboard bin itself weighed more than what was inside. H&M notes that their take-back Garment Recycling scheme, the world's largest operation, collected

18,800 tons of clothing in 2020.[21] This, they say, is equivalent to 94 million t-shirts. It's also a lousy 0.0188 percent of all textiles thrown out yearly. Moreover, all those t-shirts only equal about 3 percent of the three billion-plus garments H&M makes each year. Not something to be gloating about if you ask me.

Globally, we only recycle about 14 percent of all textile material.[22]

What happens to the other 86 percent, you ask? It will most likely end up in a landfill, probably somewhere in the developing world. Like plastics, these materials take quite a bit of time to decompose. Some textiles, like leather, take around half a century. Others, including most of what you're wearing right now, can take 200-plus years.[23] As they decompose, they leach dyes and other harmful materials into the ground. In that sense, their legacy lasts long after the material is gone.

But what if stuff does actually make it into the recycling bin? What then?

Recycling clothing is markedly different than your run-of-the-mill aluminum can recycling process. That's because there are a number of materials in each piece of clothing. You have the labels attached to the cotton shirt with thread. Jeans contain denim, elastane, and polyester, not to mention the rivets, zips, and glitter. While recyclers can separate some of this into component parts, most work is laborious, time intensive, and expensive. As we've seen, the fast fashion industry is all about keeping costs as low as possible. Do you honestly think they're investing in the proper separation of materials?

What's actually happening to those clothes intended for recycling, then? Companies are not necessarily turning your shirt into another shirt. About 1 percent goes through a process known as material-to-material recycling: 'Old wool jumpers, for example, can be turned into carpets, cashmere can be recycled into suits.'[24] Other material is downcycled and turned into blends used outside the fashion industry.

Most likely, your old clothes are destined for a long process of factory disintegration. This multi-step process requires the use of environmentally harmful chemicals. It also uses up even more water to break down textiles. So, your good intention ends up being even more costly for the planet. If anything, you're better off taking your old clothes to a charity shop or consignment store. There are also consignment websites that are worth a look.[25] At least that way, your clothes are more likely to find a new home with minimal impact. But even these options have their own sets of issues, so consider these recommendations a bandaid until we fix the larger problem.

And how do we fix the more significant issue of resource use in the fast fashion industry? The Ellen MacArthur Foundation, the world's go-to group on sustainable resourcing, outlines four ways the sector can be better.[26] The first is to rethink the incentives behind making and purchasing clothes. Next is to design products that will last more than just a week or two. We also need better integration of global supply chains and local resourcing. The last point is to create, utilise, and promote circular business models in the industry.

A quick backgrounder on circularity if you're not familiar with the concept. This practical business model accounts for a product's end of life during the research and development (R&D) phase. It's different from the typical 'make, take, waste' model we've become accustomed to. Instead, a product is designed to re-enter the supply chain instead of ending up in a landfill. This closed-loop system dramatically reduces the need for virgin materials.

The Zara model of fast fashion is antithetical to circularity. That's why it's hard to believe the laudatory greenwashing claims from them and their peers. For example, Zara and Shein have recently launched buy-back or repair programs for their clothing. Reading their press statements, it's like these moves solve all the problems the companies created. Just drop off your gently used garment, and voila!

But those in the fashion and textile industry who have tried to embed circularity into their operations, I mean genuinely embed, have a different take. Patagonia is probably the most well-known company attempting a circular business model. Their Common Threads

Garment Recycling Program was one of the industry's oldest and most closed-loop system. The company would collect used polyester base layers and then recycle them into new base layers. But even with concerted effort, they still ran into issues of economics, logistics, and scale that cut the program short.

Of course, I get these mega-brands are doing more than just putting up a few recycling bins. Many companies, especially sports shoe brands, are looking at ways to add circular design into their R&D. Adidas, for example, has partnered with Parley for the Oceans to create shoes made of recycled ocean plastic waste. Any of these are still decades away from impactful scale, though. Ciara Cates, Patagonia's lead material developer, notes just how low the bar is for the industry.

> Two decades later, even with savvy customers understanding upcycling, even with companies touting their eco-attributes, even with better infrastructure in place, circularity still doesn't figure into most of the clothing industry—not even here at Patagonia.[27]

That's why it's pretty hard to fathom a way for these fast fashion giants to meaningfully offset their dramatic resource use as things stand. We've only just scratched the surface when it comes to the industry's negative impact. From the sprawling cotton fields of western China to jute plantations in India, it takes a wide range of resources to create fabrics. The supply chain logistics of getting things from here to there contribute to even greater pollution levels. And the psychological push to buy more, more often, continues. As these companies begin to butt up against the limits of what the Earth can provide, it will take much more than their paltry measures to impact the planet positively.

Labour

According to experts, the fast fashion giant Zara churns out nearly 500 new designs each week. That's close to 26,000 in a given year! Annually, the company produces over 450 million pieces of clothing.[28] In fact, the term 'fast fashion' was coined by *The New York Times* to describe Zara's mission and production process.

When you look at any analysis of Zara's operations, you'll read congratulatory rhetoric about how streamlined their production is. Business school students study cases by experts like Donald Sull and Stefano Turconi. In their piece, *Fast Fashion Lessons*,[29] the authors laud Zara's innovative supply chain model: 'Zara maintained in-house only the capital-intensive, yet complicated, operations (like computer-guided fabric cutting); meanwhile, it outsourced the labour-intensive operations (such as garment sewing) to a network of local subcontractors, many of which were organised as seamstress cooperatives in Galicia.' They praise what they call shared-situation awareness; a fancy way of saying Zara's designers use consumer trend analysis to predict what trends are on the horizon. In short, they call Zara a model for the industry.

Wrapping all this up in jargon and with a clear dollars-and-cents lens doesn't account for the entire picture. Items don't just magically make their way from paper to peg. It takes a massive amount of environmental and human resources to make it happen. It's laughable the authors thought small seamstress cooperatives could produce all these pieces. Fulfilling Ortega's dream of cheap and fast takes much more than an innovative software program.

To get hundreds of new designs each week at rock-bottom prices means someone has to pay.

Unfortunately, that someone is most likely the person who put their blood, sweat, and tears into your clothes. According to Refinery29, a single Inditex factory (Zara's parent company) in Tunisia pushes out 150 pieces an hour. To get there,

... each worker is timed (there is a woman with a stopwatch to make sure things are running smoothly), and it's called 'working to the minute,' which means it should take 38 minutes to finish one shirt; if it takes longer than that, the plant begins to lose money. ... Employees who perform well will earn a 45-euro bonus at the end of the year. [30]

While Zara and H&M may have been the pioneers, they are hardly the only kids on the block. Names like Primark, Forever 21, and Uniqlo are just a few of the additions to a long list of fast fashion retailers. Holding companies, including ASOS, PVH, and Inditex, rake in billions of dollars a year.

In the immediate aftermath of the Rana Plaza tragedy, governments and brands expressed varying degrees of concern and outrage. Most big global brands were quick to distance themselves, making sure the public knew they had no contracts with Rana Plaza. The typical response was something along the lines of 'we're good for the economy—blah, blah, blah—we know there are problems, but Bangladesh is a developing country—blah, blah, blah—we're working to improve conditions in our factories'. What all these brands seem to agree on is that, in the words of an H&M sustainability lead, the '... best way for the country is for brands like H&M to stay there.'[31]

Anyone assuming Rana Plaza would serve as some wake-up call, or impetus for brands to improve the country, will be disappointed. There was initial progress, including creating a quasi-binding accord to help improve worker safety: 'It led to nearly 200 factories being disqualified for poor safety standards after more than 38,000 inspections at facilities covering two million workers. More than 120,000 fire, building and electrical hazards were fixed.'[32] This was superseded by a new accord that is not legally binding, calling into question the future of safety in Bangladeshi factories. There are also thousands of local factories that have no western clients. These still lack proper oversight on worker safety and labour standards.

But this isn't to say it's all bad. Around the world, many companies have made concerted efforts to improve how they produce the goods they sell.

Think back to Walmart China's Women in Factories Program detailed in Chapter 4. The program was designed to upskill female workers across Walmart's China-based factories, covering topics such as on-the-job training, business and management skills, communications, and the like. The program would also cover more of the soft skills needed to be a good worker: family planning; personal finance; health, and wellness. These were all areas most women in China just didn't learn in school.

Remember, it's not just Walmart, either. There are far more prolific programs in the fast-moving consumer goods sector that positively impact the lives of workers worldwide. The Gap's P.A.C.E. program, The HER Project sponsored by corporations like Disney, HP, and Levi's, and Plan W from Diageo are all making a huge difference for millions of workers in the developing world.

What will you do with all this new inside information? Now that you know about these programs, and are aware of what goes on inside factories, you have the power to make a more informed choice. That choice won't cost you more since the programs are already going on. Spending your hard-earned money where it can truly have an impact, though, can reverberate throughout global supply chains and the fast-fashion industry.

You probably started this section with many preconceived notions about fast fashion. I'm sorry to have to confirm them. This industry is definitely one of the world's worst when it comes to sustainability. My goal in laboriously dissecting the fast fashion industry, as with much of this book, isn't necessarily to make you do a U-turn on your behaviours. Even I still shop at some of the stores I just tore apart. Not only are they convenient, but those rock-bottom prices are hard to resist. As you'll see over and over, I don't adhere to the view that we should ditch our smartphones, go off the grid, and live on a kibbutz.

The likelihood of you buying a sewing machine and going it alone like mee-maw used to is probably pretty farfetched. My approach is much more measured and realistic.

Why didn't I include fast fashion in the chapter with tobacco, defence, and oil if it's so bad? It was a tough choice. I'll be honest. But my thinking holds that fast fashion still has plenty of opportunities to change. We've seen a few of what are probably dozens of examples where this is the case. There's no reason to believe that giants like H&M and Zara can't start to do the same.

But until they do, how are you supposed to offset all their negative impact?

First, reprioritise your buying habits. Do you really need another dress? How many pairs of shoes can one person wear? You don't even go to the gym, so why are you buying leggings from Lululemon? We learned brands are experts at using psychology to their benefit. Pushing out more, faster, triggers a primal urge in our brains to spend, spend, spend. Instead, I'd argue for a return to fashion that focuses on quality over quantity. While I recognise a lot of readers won't be able to afford the latest couture from Chanel, plenty of other mid-range options are available that won't break the bank. Think of it this way: you're investing in something that will outlast any seasonal splurges from Shein.

Next, don't believe the hype. Ultimately, consumers should be aware of what they're purchasing to make better-informed decisions. The greenwashing happening in the fast fashion industry, especially if you don't know what you're looking for, takes away that authority over your decision-making. That's how big box retailers can put your mind at ease simply by sticking a 'recycle your old clothes here' bin next to the register. They know you won't haul your clothes back to the store. You know you won't haul your clothes back to the store. But hey, it shows they care, right?

Wrong.

Now, you know the ten cents it costs to erect a cardboard recycling bin does nothing to offset the incalculable environmental impact that purchase is having.

Finally, rethink your relationship with clothes. When you look in the fitting room mirror, ruminating at length over the red dress or the blue dress, you now know what lies beyond the looking glass in some faraway factory. Clothing is more than just pieces of fabric sewn together. Not to wax poetic, but every piece tells a story. That story is written in the sweat of the people who made your garment, no matter the cost. Some may sit in a Parisian atelier, but I'd guess most live below the poverty line in the developing world. I'd encourage you to be more self-reflective the next time a cute woollen sweater goes on sale (that last point's pretty much for me).

Armed with this knowledge and by wresting back that authority from companies trying to dupe you, the power is yours. Reaching a critical mass forces companies to change. When they realise consumers are pushing back against greenwashing claims and speaking with their wallets, intelligent executives will begin to do things differently.

None of this is to guilt you. I hope that after reading the rants over the past few pages, you're now able to take a more thoughtful approach to the things you buy. If all of this does change your behaviour, even in the slightest way, then I've done my job. At least now you can't say that you didn't know.

Chapter 9

Perfect Be Damned

Throughout the ages, writers much brighter than myself have opined on the nature of perfection. Voltaire famously said, 'The best is the enemy of the good.' Shakespeare's King Lear considers that 'striving to better, oft we mar what's well'. Even as far back as the Chou Dynasty, we have Confucius encouraging his followers that it's better to be 'a diamond with a flaw than a pebble without'. In the modern business world, we often paraphrase all these by saying, 'Don't let perfect get in the way of good enough.'

This simple idea can help guide us through much of our life. It's especially helpful in framing how we should view and interact with the corporate sector. Are they perfect when it comes to positive impact? Hell no. Are they good enough to get us where we want to go? For sure.

My firm belief is that the corporate sector, more so than government or individuals, will be the ones that create lasting, sustainable change for us all. Instead of fighting and vilifying them, we should be turning to and working with the private sector to save the world. Why? Because the private sector—major businesses—has the capital, access, and capacity of scale governments, non-profits, and most private citizens simply do not.

Think about the private sector and its impact on our daily lives. There is nowhere you go and nothing you do as a modern consumer outside the purview of the private sector. They provide goods for you. They feed you. They employ you. And you love them for it. This idea that humanity can do without them, returning to the Dark Ages without mobile phones or international travel, is naïve. Our reality, our starting point, is right here and right now. Since humanity isn't going backward, how do we move ahead?

We start with knowledge.

That's been the entire point of this section: to arm you with the knowledge of when greenwashing is happening, why it's happening, and what to do about it. Our starting point was the worst of the worst— the Un-Sustainable. These companies can never be sustainable, no matter how much great advertising they put out. In particular, we talked about the oil, tobacco, and defence industries. They base their entire business model on killing people and destroying things. Unless that changes (which it won't), you should never look at them and think any of their green claims are true.

On the opposite side of the spectrum are those companies far more trustworthy than their peers. Trust comes from a combination of four things: credibility; reliability; intimacy; and self-orientation. While used primarily for interpersonal relationships, this Trust Quotient applies just as readily to corporations. I broke down the twenty-two major industry buckets and identified three sectors that outperformed on sustainability. Although they certainly aren't perfect, the pharmaceutical, medical technology, and grocery sectors do better than most. This is because they are often highly regulated and have lower levels of self-interest.

But these are companies that are trustworthy today. What should you look for as companies evolve, change, and become greener? Number one is their credibility. Remember, companies start their sustainability journey at different times in different ways. How mature is a company when it comes to sustainability and are they performing in line with that maturity? Two, look at reliability. What's their reputation, how do they handle crises, and has that reputation improved over time? Third, find out why a company became sustainable in the first place. While the ends often justify the means, sometimes it takes a disaster to get companies to become sustainable. Those organisations where a PR nightmare immediately pre-dated change are less likely to engage in all-out greenwashing.

Then we explored how a bit of innovative thinking can help the bulk of corporations genuinely change for the better. We had to step back and see differences across sustainability to frame this. There are different levels of development between companies and even within companies. Sustainability is also multifaceted, with most elements placed under environmental, social, or governance aspects. Through the lens of ESG, we read cases of companies taking a new approach to sustainability and who can serve as examples to others. Those included DocuSign's reimagination of the paper industry, how Walmart is helping their female workers in the developing world, and why more robust, diverse boards lead to better sustainability outcomes.

Finally, it was essential to call out one industry in particular: fast fashion. The modern poster child of greenwashing, the segment comes with a lot of baggage, which I wanted to dispel. We looked through the history of the fashion industry, from antiquity to today, and how these fast fashion behemoths came into existence. A discussion on transactional utility, the pleasure you get from buying something isn't just in the product itself but also in the bargain that might come with it, explained why we love Zara and H&M. Then, placing today's most prominent players in context helped frame some of the potential solutions to the sector's inherent issues.

First, reprioritise your buying habits. You don't need that extra piece of clothing, do you? Don't let their psychological tricks burn a hole in your wallet. Instead, return to fashion that focuses on quality over quantity. Second, regain your decision-making authority by being better informed. Use any inkling of greenwashing as a red flag to help you, being especially cautious of anything that seems like a bandaid to a bigger problem. Lastly, rethink your relationship with clothes. Clothing is more than just fabric and thread. There are dozens of people involved in the manufacture of one garment. Think about them the next time you're in the dressing room.

I'd argue it's high time to bring the corporate world onside in the fight for a better future. How do you, dear reader, go about doing that? The one thing all of us can do is speak with our wallets. If executives are going to be judged on financial performance, quarter by quarter, then why not hit them where it hurts? Remember, this isn't just for companies you don't like. If you have a favourite company and see them doing something you disagree with, force that change. It'll take a bit of research to find which companies are doing good for the planet and their people while also aligning with your values. Those companies that don't tick all the boxes don't deserve a single cent of your hard-earned money.

For those of us more on the front lines or who want to join the vanguard, we've got to take a hard look at how we spend our time fighting. Are we pushing shit uphill by trying to save an industry that doesn't want or deserve saving, like tobacco or defence? I know I'd much rather spend my time working with those companies that want to change, truly change, by offering innovative products and solutions to the world. Our precious energy, expertise, and sanity are not unlimited resources. We've got to spend them wisely if we want any chance of having a positive impact.

Underpinning everything, of course, is vigilance against unscrupulous behaviour. Sure, we're wrapping this up under the umbrella of greenwashing. But greenwashing is an insidious little pest. Like the iceberg that sunk the *Titanic*, the part we see only represents a small portion of what's lurking beneath the surface.

I suppose that's also the positive side of greenwashing. Once you know how to identify it, it acts like a stereogram. These picture puzzles occupied far too much of my time as a kid. You stare at a canvas that looks like a jumbled mess of shapes. If you focus long enough, a three-dimensional image pops up. Once you realise a company is lying to you in one respect, it's very easy to see how this permeates the whole organisation. You can then take this information and run from it, which is excellent. You can also take that information and use it to pressure these companies to change.

But don't go running out the door to change the world just yet. We have so much more to cover. It's not just the corporate sector that engages in dishonest greenwashing. As you'll see in the following sections, the public sector isn't much better. Government, by and large, is more concerned with saving its own hide than having our best interests in mind. Later on, we'll also examine why many individuals, while they might mean well, just don't know what they're doing.

PART 3

STATE-SPONSORED GREENWASHING

Chapter 10

Paris, *je t'aime*

Ah, Paris.

The city of light.

The city of love.

It's late on a cold December evening, but we find ourselves in an uncomfortably warm meeting room at the *Paris Le Bourget Parc d'Expositions*. Fuelled by the last dregs of urn coffee and stale croissants, men in bespoke suits and women with colourful scarves flit about the hall. Their eyes are agog. Their voices tense. Their brogues and heels are slightly scuffed from all the running around they've been doing. Above all the noise, a group of officials march onto the previously empty stage. Delegates to the twenty-first Conference of the Parties to the United Nations Framework Convention on Climate Change quiet down and find their seats.

With a bang of the felted gavel and a few words by French foreign minister Laurent Fabius, it was all over. All but two countries signed the Paris Agreement—humanity's last hope at preventing a worst-case climate-change scenario. After weeks of fraught negotiations, and years of delay, collective humanity won out over individual national interests. There would undoubtedly be much work to do, but that was tomorrow's problem. Tonight was for celebration.

The room erupted. You had John Kerry hugging his Chinese counterpart. Heads of delegations wept. There is even the famous photo of the UN Secretary-General, French President, and UN Climate Chief triumphantly raising their clenched hands. Outside, thousands of delegates cheered. Around the world, people watched the final moments play out on YouTube. I remember seeing it all from the comfort of my bed in Shanghai. Going to sleep that night, it seemed a brighter future lay ahead.

Fast forward to today, and people aren't celebrating anymore.

Since that momentous night in 2015, the world has done very little to put into practice the lofty ambitions of the Paris Agreement. Humanity has suffered increasingly hostile threats to our survival. Each year, temperatures around the globe break new records. Each year, we see stronger and stronger weather systems pulverise coastal communities. Each year, hundreds of species go extinct, thousands of people die unnecessarily, and millions of dollars flood into the coffers of the world's most polluting industries.

Paris was supposed to slow down the dystopian trajectory we were on. If that's true, I'd hate to see where we'd be today without it. Thinking about the whole situation critically, though, it isn't true. A piece of paper, no matter how lengthy and how revered, can't be the single thing that will save us. Nor can a single country, industry, or individual. The solutions to our collective problems require a collective effort. Remember this as we progress through the following few chapters.

Because now, I want to turn away from the corporate sector and look at greenwashing through a very different lens. Although we primarily associate greenwashing with large private sector corporations, they are definitely not the only guilty party. In this next big section of our discussion, we'll explore how the public sector— government, international organisations, and non-governmental bodies—are just as complicit as any oil, fast fashion, or aviation firm. While they stick very closely to the typical definition of greenwashing, you'll soon see how their brand also has unique selling points.

I'd argue the most significant difference is the excessive trust we place in these organisations to get the job done. Just like how many saw the Paris Agreement as a be-all-end-all solution to the climate crisis, it's become commonplace to rely on groups like the UN to do everything for us. That bred laziness and complacency among humanity, and hubris among these select few groups. Yes, we've contributed to this greenwashing. But these public sector institutions have done nothing to correct our viewpoints either. Perhaps they believe their own hype.

We'll start our exploration with the complex world of the United Nations and its fifty-year fight to build a more sustainable future. Although this is the big daddy of them all, it's probably the one most of you have the least insight into. As someone who's worked on the inside, I'll peel back the curtain a bit to (hopefully) make things clearer. That includes diving into the alphabet soup of mind-boggling acronyms sitting under the UN's banner and all the protocols and resolutions related to saving the planet. While Paris may be the most noteworthy, one should know about over thirty similar conferences. From Kyoto to Montreal, Doha to Rio to Glasgow, concerned environmentalists have met annually since the late-1980s. Yet, after so many decades and lots of paperwork, why are things worsening? Our ultimate task will be figuring out whether these UN conferences are still fit for purpose or, in the words of COP-26 president Alok Sharma, a 'monstrous act of self-harm'.

While the United Nations may host these conferences, the individual member states are the guests of honour. In Paris, they came from nearly every country on Earth. Heads of state, parliamentarians, diplomats, and scientists all in the same room with a shared goal. The world's best and brightest, tasked with keeping their citizens safe, have also failed. Except for reducing the hole in the ozone layer, all these leaders have yet to produce any meaningful results. In our second section, we'll examine how they engage in state-sponsored greenwashing. This type of greenwashing averts our attention from the fact they aren't doing much at all. From Saudi Arabia's ambitious green city of the future, built with oil money, to mining companies getting away with destroying some of Australia's oldest Indigenous sites, there is a lot of distraction necessary.

Then there are other hyper-national institutions that may not necessarily be political but influence a lot of the policy and direction of the world. Chief among these are large meetings of the elite, like Davos, where sustainability is often top of the agenda. However, what ends up coming out of these meetings is a lot of talking and minimal action. Sporting federations, including FIFA and the International Olympic Committee, use the façade of sport to better humanity. Meanwhile, they gladly accept billions in funding from the dirtiest of donors. Instead of using a massive platform of avid fans worldwide for good, they've become a mouthpiece for corporate greenwashing. This money is insidious and also influences some of our most prized cultural institutions. From museums to global charities, it seems there are no ends to how far greenwashing can stretch.

The performance of people in the public space—those charged with bettering humanity—would get them fired from any job in the private sector. They've failed us. Yet, you'd think they were twenty-first century saviours the way they carry on. It'll take more than the UN's triumphant pieces of paper, misallocated money to fund a dictator's pet project, or the upper echelon's obsession with disruption to building a better future. But if we keep putting our faith in these institutions, they'll keep getting away with murder (*our* murder, if we're being honest). That's why they should be held even more liable than companies for the greenwashing they commit. I guess, though, it comes with the territory. These are politicians and the elite, after all. They're experts at saying one thing and doing something entirely different. Let's see how.

Chapter 11

COP Out

'John. I need to see you in my office right away.'

These are never the words you want to hear from any boss. Coming from a six-foot-three former Finish Ambassador makes them all the more intimidating. Off I scurried past the rows of powder-blue cubicles and printing stations. Walking into Jarmo's office, I barely had time to admire the fantastic view of the Chrysler Building before he asked me to sit down.

'I'm sending you to Istanbul. There's a climate conference in a few weeks, and we need more people to run the Secretariat. Talk to Maria, and she'll get you sorted with everything you need.'

I'm sure I looked like an absolute idiot trying to contain my excitement. It was my first overseas posting, my first time meeting colleagues outside New York, and my first time working on climate. It was also a brief moment that would change the whole trajectory of my career. Until then, my work at the United Nations mainly focused on reporting, peacebuilding, and decolonisation. Little did I know my experience in Turkey (now Türkiye) would light two fires inside me: a love for travel; and a passion for saving the planet.

There was so much to do! I had to coordinate with the UN's travel agent, find a way to get out of my grad classes for a few weeks, and bring myself up to speed with the who, what, and why of the conference. The biggest hurdle, though, would be getting my coveted *laissez-passer*—that's an official diplomatic passport issued by the United Nations. Besides memories and a few grey hairs, it's probably the only thing I still have from my time at the UN. Coincidentally, I was also taking the US State Department exam around this same time. To say I had a lot on my plate would be an understatement. If I remember right, I actually came back from DC in the morning and had to leave for Istanbul that night. I was a young professional, hungry and still full of all that youth wasted on the young. Sleep could wait.

Officially, I was working inside the Secretariat of the UN Convention to Combat Desertification's Seventh Session of the Committee for the Review of the Implementation of the Convention (CRIC 7) and First Special Session of the Committee on Science and Technology (CST S-1). Yeah, that was really the name.

This was different from the big meetings I was used to. Back in New York, hundreds of delegates from all over the world would spend weeks discussing, wordsmithing, and voting on resolutions. They would take the floor and speak at length while saying nothing at all. Broad generalities, platitudes, and diplomatic niceties were the order of the day.

Here in Istanbul, it was a different affair. Of course, there were the national delegates and lengthy addresses. But there were also subject-matter experts like scientists and environmentalists in the room. You had civil society representatives having their say instead of being relegated to the back of the hall. Over the next two weeks, all would have to assess how well countries were keeping up with their commitments to combat desertification.

Let me tell you; assess they did. Instead of long, self-congratulatory speeches, delegates made remarks full of statistics and performance metrics. Some even put all their cards on the table and highlighted where they needed to improve. Experts provided helpful case studies on innovations to help push back the desert in areas where this was a growing problem. There was no fighting between delegations (although I'm sure this happened a bit in the corridors). Instead, a collegial environment permeated the conference.

That environment helped grease the wheels of negotiations. By the end of the two-week session, delegates agreed on a massive ten-year strategic plan to send...

> ... a strong message that they no longer wanted what they already had; they wanted change. Tired of making limited progress in attracting attention and resources to the Convention, parties adopted the Strategy in an effort to raise the profile of the UN Convention to Combat Desertification (UNCCD) and chart a new direction.[1]

That new direction would include a stronger focus on accountability through results-based management and the renewed importance of science over politics to guide strategy.

Has it all worked? It seems so. The Convention Secretariat (those are the people who run the show) has brought countries together to make a significant impact in pushing back desert sands. The Great Green Wall Initiative has been instrumental in building a protective barrier along Africa's Sahel region to keep the Sahara Desert in check. There is a US$300 million fund to build technologies to help with next-generation land use. The Secretariat also created easy-to-implement toolkits to assist national and local governments and civil society in preparing for, addressing, and recovering from drought.

This work was just one small piece in the broader architecture of saving the planet. While a win in this instance, I knew there were much bigger fish floating around this sea. Over the past decade, many of you have probably become familiar with them as well. But while you might know a bit more, especially if you've been watching the news, do you really know what's going on behind the scenes? Toss aside all the glitz, glamour, and big-name sponsors for some of these major environmental conferences, and what are you left with?

Some successes.

Some failures.

And a whole lot of greenwashing.

In the weird, complex world of international diplomacy, the United Nations is unique in its ravenous thirst for acronyms. My first job in the organisation was as an English-language editor in the DGACM—Department for General Assembly and Conference Management. I encountered dozens of these daily. They included the obvious: UNSC for United Nations Security Council; PKO for peacekeeping operations; and SG for Secretary-General. Then you got into those used as shorthand for formal country names, like UKGBNI for the United Kingdom of Great Britain and Northern Ireland, and DPRK for the Democratic People's Republic of Korea. That's North Korea. South Korea is simply the ROK. Big blocks of countries get their own acronyms, too. You've got the G7, G8, G10, and G20. The G77 + China is primarily developing countries. CANZ is short for the Canadian, American, Australian, and New Zealand alliance. BASIC and BRICs include Brazil, Russia, and India. SIDS are small-island developing states.

My favourites were always the acronyms for and related to peacekeeping operations. These would typically have extremely long names, leading to some pretty crazy outcomes. MINURSO is a mission in Western Sahara, while MONUC is another in the DRC (Democratic Republic of the Congo). MOSS stands for minimum operational security standards, and NEPAD is the New Economic Partnership for Africa's Development. FANCI are the *Forces Armées Nationales de Côte d'Ivoire*. WIDER is the World Institute for Development Economics Research. MANPADS, though, has to take the cake—MANPADS is short for a type of surface-to-air missile called a man-portable air-defence system.

Are we having fun yet? If so, the acronym cavalcade doesn't stop when we get to the UN's work saving the environment. When it comes to sustainability, the big daddy is the United Nations Environment Programme, or UNEP.[2] Established in the early 1970s, the UNEP's mission has been to coordinate all the then-disparate efforts across the UN system. It did so for a while. During this period, UNEP focused primarily on science and policy related to climate change, assisting in disaster scenarios and providing technical manuals related to climate resilience. Its work put it well ahead of other scientific bodies in predicting the impacts of climate change. In a sad bit of irony, it was a vocal chorus of countries from the developing world speaking out against UNEP's work. Why worry about the environment when there were more pressing economic matters? Obviously, this has come back to seriously bite them on the ass.

Following a decade-plus of leading work, UNEP began creating what marketers call a house of brands. They served as a parent organisation, but what was once a coordinating body began to break up into disparate parts. The first significant shift was the creation of the IPCC, the Intergovernmental Panel on Climate Change, in 1988. Unlike UNEP, scientists (and not career diplomats) run the IPCC. They spend much of their time collecting, collating, and reviewing all the world's existing scientific knowledge about climate change. They take all this and provide advice to governments as well as annual reports on the state of the climate. These annual reports have begun to make headlines as the findings grow increasingly dire, perfect material for the doom-tastic ten o'clock local news. The IPCC is widely considered the most trusted source of climate change scientific information, even garnering a Nobel Peace Prize in 2007.

Like Cain slew Abel, the IPCC eventually dwarfed the work of its parent organisation and called into question the usefulness of the UNEP. Following the IPCC's damning *Fourth Assessment Report*, published in 2007, forty-six countries called for UNEP's complete dissolution. Allegations of corruption and a far too limited scope of powers were the nails in the proverbial coffin. In its stead, supporters cited the need for a more powerful body akin to the World Health Organization. Led by then-French President Jacques Chirac, the Paris Appeal (not to be confused with the Paris Agreement, which we'll get to) urges 'massive international action to face the environmental crisis'.[3]

Chirac's vision never materialised. He left office only a few months after this declaration, and without a vocal champion, the idea of a UN Environment Organization fizzled out. Calls for UNEP reform, though, have not.

The danger in creating a house of brands, even inadvertently like UNEP did, is that you risk your sub-brands becoming more recognisable, popular, and successful. You might not think, *I want to buy something from LVMH today*. Many would, however, think about buying from Tiffany, Sephora, or Kenzo. AB inBev might not whet your appetite, but having an ice-cold Corona or Stella Artois certainly will. The same has happened with UNEP. While people know about it, one hears far more about the IPCC and some of the other sustainability work done by the UN, to which we now turn.

The next major acronym you need to know is UNFCCC, the United Nations Framework Convention on Climate Change. Even though the UN Environment Programme had been in operation for two decades, there was no international mechanism to get anything done. The UNFCCC was the first treaty to provide this framework. At its core, the UNFCCC keeps all nations on track to reduce greenhouse gas emissions and mitigate humanity's negative impact on the climate. The Convention divides countries into three groups. Basically, it's those who've caused the problem (developed countries), those who are feeling the impacts of the problem (developing countries), and those who can pay to fix it.

The UNFCCC established an annual meeting to gauge progress against all the Convention set out to do. This meeting is an acronym most of you are probably familiar with because it's the one that gets the biggest splash in the media: COP. That stands for a very unhelpful descriptor: Conference of the Parties. The 'Parties' are the nearly 200 countries that have signed the UNFCCC. The rest is pretty self-explanatory. Each year they fly in on private jets to discuss the Convention, see how well we're all doing (spoiler alert: we're not), and pat each other on the back. Unlike my time in Istanbul, COP has evolved to become performative nonsense on a global scale. But as we'll see shortly, this is a new development. COP has a history of outstanding accomplishments pushing forward the global sustainability agenda.

Maybe you haven't heard about COP specifically, the UNFCCC, or the UNEP. But you may know some cities that have hosted big environmental conferences. Don't worry. There's no need to know over thirty different conferences. Instead, focusing on the ones with the most significant impact is perfectly fine. You've got Rio, Kyoto, Copenhagen, and Paris. Arguably, the most important of these was the 2015 Paris meeting which led to the landmark Paris Agreement. But as discussed earlier, this was just a culmination of nearly two decades of climate work from politicians, scientists, and civil society. A quick breakdown should help make sense of things.

Let's start with Montreal in 1987, a time before COP was even a glimmer in its mama's eye. Here, global leaders signed the Montreal Protocol. The Protocol was the key agreement to combat the growing hole in the ozone layer over the Antarctic. Montreal put restrictions on aerosols containing an alphabet soup of chemicals. It's why you don't see things like chlorofluorocarbons, hydrochlorofluorocarbons, or hydrofluorocarbons in your favourite products. The Montreal Protocol was also the first major agreement relating to protecting the environment. In essence, it jump-started us along the COP trajectory.

Next came the Rio Summit in 1992. Often dubbed simply the Earth Summit, Rio was KM Zero for all future meetings. Today you'll often see conferences, resolutions, or speeches using Rio + [insert number here]. There's Rio+10, Rio+20, and Rio+30. By far, the biggest outcome of the Rio Conference was the establishment of the United Nations Framework Convention on Climate Change, the UNFCCC, which we've just explored. Other important things happened, too. The Summit agreed to a Convention on Biological Diversity, put a stop(ish) to development on indigenous lands, and adopted the infamous Agenda 21.

Originally envisaged as a way to embed sustainable development at all levels of government, Agenda 21 soon became the wet dream of conspiracy theorists around the world. So began the stories of the UN trying to control populations and create a new world order. Never mind that the UN can hardly organise a picnic. No, these whack balls insisted Agenda 21 was using environmentalism to disguise much more sinister motivations.

Tea Party members in the US (remember them?) claimed Agenda 21 was a UN plot to urbanise the United States. Ultra-conservative groups like the John Birch Society saw Agenda 21 as eco-totalitarianism. Even that bumbling idiot Glenn Beck wanted to have a say. He wrote an entire book trying to discredit Agenda 21, using a poorly described dystopian future if the UN got its way. I mean, you've got to see the intro to believe it.

> Emmeline and her family live in a place that just a generation ago was called America. Now, it's simply known as 'the Republic'. There is no president. No Congress. No Supreme Court. No freedom.
>
> There are only the Authorities.
>
> This bleak and barren existence is all that eighteen-year-old Emmeline has ever known. She dutifully walks her energy board daily and accepts all male pairings assigned to her by the Authorities. Like most citizens, she keeps her head down and her eyes closed.
>
> Until the day they come for the one she loves most.[4]

If I were Margaret Atwood, I'd be on the phone to my lawyer.

Without a basis in fact or reality, Beck claims Agenda 21 will wipe out 85 percent of the world's population, create a police state, and give the UN total control over everything. In reality, Agenda 21 mainly created a few funds, set the basis for the Sustainable Development Goals, and established the Convention to Combat Desertification (those folks I worked with in Istanbul). You might laugh, but Beck's 2012 book currently has a 4.5-star rating on Amazon and a 4-star rating on Goodreads. People are listening, and they believe this junk, no matter how ridiculous it truly is.

Even with the weight of the world behind them, agreements take time to come into force fully. In the case of the UNFCCC, official adoption didn't happen until two years after Rio in March 1994. Soon after, the UN held its first official Conference of the Parties in Berlin. In total, there have been twenty-seven annual meetings of signatories to the UNFCCC. By the time you read this book, one more will likely have taken place in the UAE.

The next big important COP came in 1998 in Kyoto. This is where the world adopted the Kyoto Protocol. Pushed mainly by then US President Bill Clinton and Vice President Al Gore, the Kyoto Protocol was the first action implementing parts of the UNFCCC. It also contained the embryo of an international emission trading scheme. While they are increasingly going out of fashion, trading schemes allow companies to buy credits that offset their carbon emissions. It's like when you can pay to offset that flight to Europe, but this is for highly polluting corporations. Overall, the Protocol aimed to reduce man-made emissions wherever possible.

Like Agenda 21, one could write books about the Protocol. While not the stuff of conspiracy theories, the Kyoto Protocol came under fire from developed nations that would ostensibly get hit with massive fines for polluting. Canada, for example, pulled out of the Protocol in 2012, citing a possible US$9 billion penalty that would destroy their economy.[5] For many of these same reasons, the US Senate never ratified accession to the Protocol even though it was primarily an American initiative. Through fits and starts, the Kyoto Protocol chugged along for nearly two decades. Without a successor, it expired in December 2020.

Fast forward to 2009 in the Danish capital, Copenhagen. This conference was supposed to be a historical highlight for the UN but ended up as one of its biggest diplomatic blunders. The goal: set out a strategy for the UNFCCC that would replace the Kyoto Protocol when it expired. Very public squabbling between developed and developing nations meant little happened. If you think I'm being flippant, read some of these headlines from the media who were there. A *Der Spiegel* headline reads, 'The US and China Joined Forces Against Europe'. The *BBC* writes, 'China Rejects UK Claims It Hindered Copenhagen Talks'. The *Australian Broadcasting Corporation* posits, 'Developing Nations 'Resisted' Climate Deal'. The *Sydney Morning Herald* claims, 'India Confesses It Helped Derail Copenhagen Deal'.

In this blame game of a conference, tempers flared and egos got manhandled. President Obama is even said to have stormed into a meeting to scream at Chinese Premier Wen Jiabao. Ultimately, the 2009 Conference didn't achieve its official aim. On the sidelines of the Conference, though, a lot of work was happening to reach what was called a political accord. The US, China, and two dozen other countries signed the accord, creating a fund for environmental resourcing. Done outside the auspices of the Conference, most people consider it a tangential win.

So with the failure of the Copenhagen Conference, the goalpost moved yet again. As the world burned, leaders met a further five times before finally coming together at the most well-known of UN climate conferences: COP 21 in Paris. It was here delegates passed the—wait for it—Paris Agreement. You might also hear it as the Paris Accord or the Paris Climate Agreement, which are all correct. If you've been paying attention over the last few pages, you understand how difficult it is to get countries to agree on anything. Yet, with the Paris

Agreement there was almost universal adoption. Only two countries, Iran and Syria, have yet to sign. That's a pretty big win on its own.

Beyond this monumental agreement, why else is Paris so noteworthy? At its core, the Agreement seeks to rev up an ageing UNFCCC in three ways. First, leaders set the famous 2-degree centigrade limit on global temperature increases. There is broad scientific consensus that going beyond this number would lead to irreversible changes to ecosystems and our way of life. Second, the Agreement sets the foundations for global climate resilience. This is a fancy way of describing all the things we must do to adapt to the impacts of climate change. Lastly, it puts in place financial mechanisms to help pay for all this.[6]

The most significant difference with Paris is how the Agreement pushes countries to participate. One of the biggest, if not *the* biggest, sticking point with the Kyoto Protocol was everything being legally binding. With Paris, actions would come from consensus and nationally agreed non-binding targets. From the Paris Conference, we also get the mother of all acronyms to describe this process: CBDRILONCWRC.[7] That indecipherable smattering of letters stands for Common But Differentiated Responsibility In Light Of National Circumstances With Respective Capability. Essentially, each nation is responsible for doing its part. How that materialises, though, will vary country by country.

That's a double-edged sword. Sure, it makes all of these climate commitments an easier pill to swallow. On the other hand, it also lacks teeth because there is no accountability mechanism. Ultimately, the Paris Agreement was a hard-fought win for environmentalists worldwide. From the Center for Climate and Energy Solutions:

> … the agreement represents a hybrid of the 'top-down' Kyoto approach and the 'bottom-up' approach of the Copenhagen and Cancun agreements. It establishes common binding procedural commitments for all countries, but leaves it to each to decide its nonbinding 'nationally determined

contribution' (NDC). The agreement establishes an enhanced transparency framework to track countries' actions, and calls on countries to strengthen their NDCs every five years.[8]

Those are some of the most prominent climate conferences you might come across. None of this is to say, though, that the other conferences were any less important. Like the pieces of a puzzle, each meeting contributed its particular part to the big picture: COP 16 in Cancun established the US$100 billion Green Climate Fund; COP 7 in Marrakech created enforcement mechanisms to fine polluters; and COP 8 in New Delhi started conversations on technology transfers from developed to developing countries. Even outside these notable conferences, rest assured that all the other UN agencies tasked with building a sustainable future are hard at work. More than anyone else, these behind-the-scenes players are lifting the heaviest loads.

Given the state of the environment, with humanity teetering on a knife's edge, it makes perfect sense for the media to run stories about major UN climate conferences. Since 2010, and especially following the Paris Conference, there has been a noticeable uptick in the amount of media devoted to climate change, the environment, and sustainability. Those on the front lines have finally got what we've been asking for: a platform. It's no longer about getting people to listen, which is a good thing. Now, our job is to figure out how to capitalise on this captive audience.

If the latest Conference of the Parties is anything to go by, we're squandering the exact thing we've been fighting for.

Sharm el-Sheikh is a gorgeous little resort town nestled against the crystal azure waters of the Red Sea. Known affectionately as 'Sharm', it's become quite a hotspot for the monied jet set to escape the brutal European and Russian winters. Egyptian families and businesspeople also flock there to get away from the hustle of Cairo's streets. Scuba diving, golfing, and shopping entertains by day. Five-star hotels, luxurious spas, and world-class restaurants entertain by night. Some have even gone so far as to describe Sharm as the perfect holiday destination.

Sharm has also become a favourite location for major international business and diplomatic conferences. Purpose-built conference centres with a capacity for thousands attract groups like the World Bank, Arab League, and even the United Nations. It is designated a UNESCO 'City of Peace' given the diplomatic work that's happened here. Against this backdrop, and the shifting sands of the Sinai desert, we find ourselves at COP-27.

Just a short month before the opening of the Sharm Conference, UNEP released a report saying there is a 96 percent probability we'll pass the critical 1.5-degree global warming threshold in the coming decades. This followed numerous reports by the IPCC and other scientific bodies warning us about the increasing threat of man-made climate change. The journal, *Earth System Dynamics*, noted humanity had already passed five major climate tipping points, which would cause a domino effect of problems.[9] For those who maybe missed these reports, you couldn't escape the heartbreaking stories making international news. Pakistan had just been hit with months of rain, causing the country's worst flooding and nearly 2000 people dead. Europeans suffered through their highest recorded summer temperatures. America was in the midst of a once-in-a-millennia megadrought.

But looking at all the glad-handing and pale, male, stale panels in Sharm El-Sheikh over those two weeks, you'd be forgiven for thinking we're in the clear. That's because COP-27, more than any of its predecessors, was a sickening exercise in greenwashing and empty platitudes. The whole thing has become an uncontrollable beast at this stage. The two conferences couldn't be any different if you think back to the start of this chapter and my time in Istanbul. Even compared to the political debacle that was Copenhagen, or non-conferences like COP 5 in Bonn, COP-27 was something else entirely.

Firstly, the foxes seem to be guarding the henhouse. While UN

Secretary-General Antonio Guterres made an impassioned speech demanding a crackdown on greenwashed net-zero commitments, what about all the greenwashing happening at COP itself? Coca-Cola, one of the world's biggest plastics polluters, was a key sponsor of the Conference. The Climate Action Tracker notes there are precisely zero countries on track to meet their Paris climate targets, yet many set up national pavilions to host lavish events. My personal favourite is Australia, which with zero sense of irony or optics, partnered with the Carbon Market Institute to run panels.

Secondly, this isn't the Grammys. The whole COP experience has turned into a circus. The conference I worked at in Istanbul was full of people who could get the job done: subject-matter experts, civil society, and a few political delegates. Looking through my LinkedIn feed during those two weeks of COP-27, I was bombarded with everyone from CEOs to investors to unknowns flying to the desert to discuss saving the planet. They're taking red-carpet photos for clout and hosting panel after panel. But is any of this going to translate into the critical action we need to be seeing?

Lastly, the presence of oil lobbyists and the fossil fuel industry was especially concerning. Greta Thunberg put it best when she asked if one would invite mosquitoes to a malaria conference. It seems COP-27 left the fly-wire screen open. *The Guardian* reported a record number of oil lobbyists attending the event.[10] What's worse is that these lobbyists outnumbered all delegations, bar one. This, coupled with pushback on language around fossil fuels by Saudi Arabia, is a worrying development signalling how entrenched these players are in the big picture.

The Conference did ultimately end with many positive outcomes. Given COP-27 was not the most conducive environment for negotiation, these wins are even more impressive. The uptick in civil society, women, and youth involvement was particularly heartening. As these groups were always present, they did also get a seat at the table in Sharm. That includes Indigenous voices long placed at the periphery of these events. We also had the monumental passage of a compensation fund for countries most impacted by climate change. This fund has been a contentious issue since my days at the UN, so it's great to see a resolution, even if it is only theoretical at this point.

But I fear history will remember COP-27 less for all of these positives and more for its negative optics. Outsiders must have scratched their heads at the rampant greenwashing, hordes of eco-influencers grabbing the perfect photo op, and hundreds of flights landing in a desert resort town. Their brains really must have melted from the cognitive dissonance of the fossil fuel lobby getting undue airtime. Delegates working behind the scenes and into the wee hours wordsmithing a document aimed at positive impact means little if the public doesn't see it happening. It means even less when the only document they do see is full of technical jargon and undecipherable UN-ese (in which, dear reader, you must now be an expert!).

Ultimately, we need to take a long look in the mirror and gauge whether the COP format is still fit for purpose. Or does its continued existence risk poisoning the well for climate action? The only way to do that is to compare COP with what's been happening globally since 1995. Granted, one can't expect an annual meeting of politicians to make a tremendous amount of impact. But with all the resolutions, accords, and agreements, you'd assume we would be moving the needle in the right direction.

You don't have to be a scientist or UN insider to know this is hardly the case. As unfortunate as it is, the climate crisis has only grown in intensity. Since 1995, global temperatures have steadily increased. Many labelled the hottest year on record lost their title only a year later. It's all but certain we'll pass the 1.5-degree threshold much sooner than expected. Some experts put that in the next two decades rather than the original estimate of 2100. Averting climate disaster is no longer the goal. Now, it's adapting to and being resilient against an Earth that's pissed off with us. In their April 2022 report,[11] the IPCC noted how to stave off even worse disasters. All we need to do is peak all global emissions by 2025 and then reduce them 43 percent by 2030. Easy-peasy.

As with COPs past, COP-27 was just another instance of running in place. Delegates kicked most of the significant decisions down the road to COP-28. But this is the same thing that's been happening since the first COP in 1995. Sustainability professionals often say we're building a better world for future generations. How much further into the future do we need to go?

It's clear a shakeup to the system is long overdue. Here's what I would do in the mythical world where I can snap my fingers and change things.

First and foremost, it's critical to recognise how overgrown and bloated these marquee events have become. They are trying to be all things to all people. The interests of politicians, business leaders, and the general public are rarely going to align cleanly. Shoehorning all of this into a two-week conference doesn't make practical sense. There's the adage that you accomplish nothing when you try to do everything. I'd argue this perfectly describes what we're witnessing.

In response, we should look to niche down as much as possible. Think about a business or school project you've worked on. Did you approach it without a plan? No. Did you try doing it all at once? Of course not. You broke it down into smaller, more manageable chunks. That way, you could tick off key milestones as you finished them. Not only did that likely make the project less of a monster to tackle, but kicking those smaller achievements in the butt gave you the energy to keep going. If all you see is the enormity of the entirety, then frustration, confusion, and burnout are bound to follow.

Back in 2019, I emceed the UN Environment Programme Finance Initiative's Asia-Pacific Roundtable. Much like the Istanbul conference I attended a decade before, this was pretty niche. Because it was so niche, you had actual experts in the room instead of random clout chasers with no understanding of the subject. Even though it was a short event, the dialogue was deep and the information-sharing broad. It was there we introduced the Principles of Responsible Banking, the world's foremost banking network. Delegates also worked through risk management for insurers, launched a monitoring report for ESG disclosure in China, and began a dialogue with investors that continues to grow. Not bad, if I say so myself. And it seems we got a lot more done than everyone at COP-27!

Second, we need to find a dragon we can all slay. Like breaking a work or school project down into bite-sized chunks, we need to rack up smaller wins to inspire us to continue. It's demoralising to see these conferences happening every year and seemingly accomplishing nothing. That's because the common enemy, climate change, is way too big. It's long-term, too, meaning we're not likely to see the results of our efforts during this lifetime. Plus, plus, climate change means different things to different people. What concerns the people of The Seychelles is not the same as someone choking on pollution in India.

But we don't need to get ourselves down and fall prey to thinking it's all too much. We don't even need to look that far back in history to find a time when having a common environmental enemy worked in our favour. Just consider the issue of the ozone layer. In the 1980s, people realised the world was getting just a bit too hot. British scientists discovered the culprit was an increasingly large hole in the protective layer of our atmosphere known as the ozone. This layer keeps out the harmful radiation from the sun, ensuring all of us can go on a walk without having to use SPF-9000 sunscreen.

Scientists soon linked the use of chlorofluorocarbons (CFCs) to ozone depletion. CFCs were part of most aerosol products at the time. You've got to understand how much this era relied on big hair. As a shock to every global consumer, the cans of hairspray and shaving cream they so readily used were destroying the planet. The media immediately began to push the story, scaring the hell out of everyone, everywhere. In one *Newsweek* interview (which probably was a little on the nose at the time, as it certainly still is today), a terrified environmentalist said the threat was like 'AIDS from the sky'.[12] The terror reached such a fever pitch the international community banded together to sign the Montreal Protocol, effectively banning the use of CFCs. In what is probably the most prominent environmental victory of our time, we have been able to claw back and fill in the ozone hole in the Antarctic. While it still ebbs and flows with the seasons, scientists believe we will entirely heal the ozone layer hole by 2050.

Why this was such a successful campaign comes down to three things. First, it was easy to understand. A hole in the sky equals increasing temperatures threatening to wipe out the Earth. Got it! Second, it was timebound. If we don't do something today, you can kiss tomorrow goodbye. Lastly, it was a problem people could easily do something about. If you want to save the planet, all you have to do is replace your aerosol can with a pump-spray alternative. Compare this to the grand issue of ending climate change. Is climate change easy to understand? Hardly. Is it timebound? Most people wouldn't think so the way we keep talking about saving our children's children, going net-zero by 2050, or X metres of sea-level rise by next century. Is climate change something Joe Blow can do anything about? Even though they might do their part to recycle or save water, I bet they think it's all kind of pointless.

Lastly, injecting a bit of accountability into all this wouldn't be a bad thing. How we accomplish that, with 192 countries getting an equal say, is anyone's guess. This isn't a new subject, either. Calls for reform go all the way back to when I was with the UN. We saw it with the Security Council when the Russian Federation continued to block discussion of its illegal invasion of Ukraine by using the veto power. It's a common talking point with the Human Rights Council, which counts some of the world's worst offending abusers as members. So, too, has it hit the United Nations Environment Programme. In 2012, the UN General Assembly passed a resolution calling for reform to the UNEP. Its primary concern: improving the governance of the organisation.

My higher education experience, from representing Guinea-Bissau as an undergraduate in Model UN to a Master's Degree in the Study of International Orgnaizations, taught me a lot about the workings of the United Nations. One critical thing all of this academic experience left out, though, isn't something most people would even consider. It wasn't until I had worked inside the belly of the beast for a while that I came to this realisation. The United Nations wasn't built to create and drive change, as many think it was. Sure, the organisation does a lot of fantastic work around the world. The list of accomplishments over its seventy-plus years has made the world much better. But all of this is circumstantial.

The real reason the world came together after a decimating half-century of conflict is right there in the opening lines of the UN Charter: 'We the peoples of the United Nations[,] determined to save succeeding generations from the scourge of war, which twice in our lifetime has brought untold sorrow to mankind ... unite our strength to maintain international peace and security.'[13] That's it. Maintain the status quo. While the Charter does contain language around advancing humanity through economic and social means, this is nondescript and buried in the text. When those delegates met in San Francisco in 1945, keeping things copasetic and stable was all they cared about.

Once I came to this realisation, it put much of my work in context. Sitting through the same meetings, year after year, seemingly accomplishing the same things. Updating the same documents by date, not by content. Even the bloody headquarters itself is a time warp. When you walk into the building off First Avenue, it's like being transported to the 1950s. The design, furniture, and colour scheme scream mid-century. Yet all of this is for a reason. Keeping things as they are makes it a lot easier to balance the ship.

To be absolutely clear, there is nothing wrong with this. I'd hate to imagine what the world would be like without the United Nations. Having been in the room discussing disarmament, peacebuilding, and decolonisation, I know it doesn't take much to piss delegates off.

Without the UN there as a referee, I doubt the world as we know it today would even exist. So the organisation deserves all the praise we can muster for accomplishing its intended mission.

But that doesn't mean it is fit for purpose when it comes to pushing for the kind of change we need to save our planet today. As we've seen, the status quo is the enemy. In trying to stay the same, we're swiftly going backwards. Perhaps the UN can change so it's more in line with the demands of the twenty-first century. Based on my experience, even if that were to happen, it would take a very long time. So then maybe there are other organisations out there that should be front-and-centre when building a sustainable future. As it stands, however, the United Nations, its environmentally focused bodies, and all the work within them haven't accomplished what we thought they should. To continue placing our trust in them is only a delusion.

Chapter 12

Toeing the Line

It's a stiflingly muggy afternoon in Bangkok. Chaos on the city's streets makes it feel even worse. Tuk-tuks, buses, and motorbikes spew exhaust everywhere. Overhead, the metro rumbles along. Every so often, you'll get a cold draft coming out of an air-conditioned mega-mall. Then an instant later, it's gone. Even though my phone tells me one temperature, the wet-bulb real feel is at least five to ten degrees warmer. An unending torrent of sweat pours down my back.

And you know what? I love it.

Even though the heat can be oppressive, Thailand is probably where I'm happiest in the world. It's a feast for the senses, offering more than can be done in one lifetime. Of course, I've got my favourite haunts (looking at you, Terminal 21). Yet every time I'm there, it's impossible not to get swept up in a new experience; a place to see, or a dish to try. Today, though, I just wanted to relax.

As I made my way from the gym to my apartment, I couldn't wait to leave the noise and heat behind. An afternoon of mango sticky rice and television was just what the doctor ordered. Plopping down on the couch, I grabbed the remote and started scrolling through some YouTube options. That's when I first saw it: an advertisement for this thing called The Line.

It was a very slick two-minute presentation by a group called NEOM. From space, we see the sunrise behind a darkened Earth and are then quickly transported to a sprawling city in the desert. A gentle female voice admonishes humanity for having lived in squalor since time immemorial. Now, though, she promised a 'revolution in civilization [sic] is taking place'.[1] Hyper-realistic graphics take that sprawling city, uproot all the buildings, and condense them into a very long, thin line stretching for kilometres across the Arabian Peninsula. Up pops a wall on either side of the buildings, encasing them in this highly reflective mirrored surface. Like an oasis from *One Thousand and One Nights*, The Line is now complete and worthy of Scheherazade herself.

That was just the first thirty seconds. But wait … there's more.

Now we're flying high above the rooftops, dodging our way through alleys, over perfectly manicured parks, and inside a maze of infrastructure. Then the stats start flying, too. The Line will be 500 metres tall, about the same as the Empire State Building, 200 metres wide, and 170 kilometres long. Yet it will only take up a minuscule 34 square-kilometres of total space. It's easy to see the difference when you compare it to places like Melbourne at nearly 10,000 square-kilometres. Even the world's most compact big cities, like Manhattan or Hong Kong, are still dozens of square kilometres bigger than The Line.

Innovative urban planning will take human needs into account. Its nine million residents will never need to leave their neighbourhoods as everything—food, work, entertainment—will be available within a five-minute walk. An ultra-modern metro system can take you from one end of The Line to the other in less than twenty minutes. AI will power all services within the city, so you can spend time doing the things that matter most. The list goes on.

Best of all, The Line positions itself as the future of civilisation because it dramatically reduces the negative environmental impact of a city. Its unique design will create a micro-climate to keep things temperate, even in the blazing Middle Eastern sun, and provide 100 percent renewable energy. There is no need for cars, trucks, or buses since you can walk to whatever you need. Thus, The Line will create zero carbon emissions. You'll also have easy access to nature should you ever consider stepping outside.

'The Line. The city that delivers new wonders for the world.'

Except it's all bullshit.

I'm actually offended by how obviously BS the whole thing is. Now instead of having a lovely relaxing afternoon, I've gone down a rabbit hole learning all about The Line, NEOM, and how one of the world's biggest polluters is trying to gaslight us all into thinking it's saving humanity. Let's break things down.

First off, follow the money. Who's funding something is usually a pretty good indicator of intent and ulterior motivations. In the case of NEOM and The Line, these are state-sponsored initiatives by Saudi Arabia. Given the country's awful track record on the environment and human rights (plus its penchant for dismembering journalists and funding terrorists flying jets), Saudi Arabia's been on a decade-long campaign to clean up its image. The face of the operation is Crown Prince Mohammed bin Salman Al Saud, heir apparent to the Saudi throne. MBS, as his buddies like to call him, isn't even forty years old. Because of this, he has a much better grasp of PR and what it takes for inclusion amongst the 'civilised' nations of the world.

That's why he's investing so much in face-saving projects like The Line. While massive on its own, The Line is just one part of the larger NEOM project. NEOM is to be a futuristic city the size of Massachusetts in the northwest of the country. It will somehow build floating islands and have year-round snow fields for skiing. All of this is part of Saudi Arabia's Vision 2030 plan, which aims to diversify and future-proof the country's oil-dependent economy. It also all comes with a hefty US$500 billion, with a B, price tag.[2] But when you're the de facto ruler of an oil-rich monarchy, that's nothing. As Mom used to say—if you have to ask how much, you probably can't afford it.

Second, pull up the covers and see what's lurking underneath. It's no secret Saudi Arabia has a thorny relationship with human rights. The country consistently ranks as one of the worst places in the world if you value little things like freedom of expression and religion, the right to privacy, as well as rights for women, girls, and migrants. Religious police enforce a draconian observance of Sharia Law for both Muslims and non-Muslims. Public whippings, hangings, and beheadings are common punishments. In March 2022, international media reported a mass execution of eighty-one men. That's why NEOM is such an essential distractive measure for MBS.

But even The Line is shining a light on the country's views on human rights. Although NEOM's proposed land is sparsely populated, it's not entirely empty. The nomadic Huwaitat Tribe have roamed this area for centuries. None of that matters to MBS. The Government is forcibly evicting around 20,000 members of the Tribe to make way for the bulldozers.[3] Anyone who gets in the way is being sentenced to death. To show you just how bad this is, I want to quote a report from human rights group ALQST at length.

Saudi Arabia's Specialised Criminal Court (SCC) has sentenced to death three members of the Huwaitat tribe whose family, along with several others, have been forcibly evicted and displaced to make way for the Neom megaproject being pursued by the Saudi authorities.

On Sunday, 2 October 2022 the SCC, the court set up to handle terrorist cases, handed down death sentences on Shadli, Ibrahim[,] and Ataullah al-Huwaiti. Shadli al-Huwaiti is the brother of Abdul Rahim al-Huwaiti, shot dead by security forces in April 2020 in his home in Al-Khariba, in the part of Tabuk province earmarked for the Neom project, after he posted videos on social media opposing the displacement of local residents to make way for the project. On 23 May 2022, Shadli went on hunger strike in protest against ill-treatment and being placed in solitary confinement, and after two weeks the Dhahban Prison administration inserted a tube into his stomach to force-feed him, a form of torture.

Ibrahim al-Huwaiti was one of the delegation of local residents who in 2020 met the official commission charged with securing government title to the lands required for the Neom project. Ataullah al-Huwaiti was also seen in several video clips talking about the misery his family and all the other displaced residents were facing as a result of the decision to evict them.[4]

Global human rights groups are calling out the international firms involved in all this, from designers, to architects, to the builders themselves. Some of the biggest names taking part in the NEOM project are Australian studio Bureau Proberts, German studio LAVA, UNStudio from The Netherlands, Morphosis and Aedas from the United States, and most unfortunate of all Zaha Hadid Architects (may she rest in peace). Amnesty International's Peter Frankental noted that while the Government may be responsible for these actions, the international firms benefiting also have blood on their hands.[5]

Finally, put on your thinking cap and critically examine the whole thing. Does it make sense? Will anyone actually want to live there? Is it sustainable? The answer to all three of these questions is a definitive no. To start, it's just not logical. Why would you build an entirely new city in the middle of the desert when major population centres already exist? If the issue is failing infrastructure, take all that money and use it to fix existing problems. If it's because people are tired of the dust and grime in Riyadh, buying a pressure washer seems a much cheaper solution. Plus, having an entire city run by autonomous robots, flying drones, and metro lines that could easily break down sounds like the stuff of nightmares.

Next, will people actually want to live there? This, of course, is anyone's guess. Elon Musk still has fanboys, so anything's possible. To uproot one's life and move 1500 kilometres is a big ask. There had better be something more than flying cars to draw me in. Proximity? While laudable, it's unrealistic to have everything within a five-minute walk. Small footprint? People need space. Cramming them into an area only 200 metres wide isn't super attractive for most. Ease of living? Maybe, but we've all seen what happens when we give AI too much power. So unless a T-800 comes back from the future to move me personally, I'm staying put.

Then the biggest question of all: is it sustainable? From a human rights perspective, we've just seen that it's nowhere close to hitting the mark. With environmental sustainability, too, all the stats, slick adverts, and money can't hide the project's inherent problems.

Specifically, there are a few significant environmental concerns with The Line. Remember, just because The Line has no cars doesn't make it green. Logistically, how in the hell are you going to create a city from nothing in the middle of nowhere without having some carbon footprint? Transporting materials during construction will have an impact. Transporting goods, people, and services when the whole project's up and running will have an impact. Construction is one of the most polluting industries in the world. A project of this scale will be pumping out record-setting amounts of pollutants now and up through its intended completion date of 2045. Plus, the massive amount of resources needed for something like this will be difficult for the Earth to recoup.

Impact on the natural environment is critical, too. I wonder if MBS did a proper environmental impact assessment before he launched this idea to the world (my guess is no). In it, the assessment probably would have told him all about how building a skyscraper-tall wall hundreds of kilometres long might disrupt migratory patterns. Or how disposing of waste from nine million people shouldn't be a footnote in a TED-style presentation. It likely would have pointed out, too, that having a massive metallic, mirrored wall could be problematic for birds, animals, and anyone without sunglasses.

Oh yeah. There's that pesky bit about everything going on *outside* The Line, as well. While The Line might position itself as sustainable, I doubt it's green enough to offset all of Saudi Arabia's oil peddling. Currently, Saudi Arabia has the world's second-largest oil reserves and is second globally in oil production.[6] It is also sixth in oil consumption, both by number of barrels and per capita use.[7] The country exports about 60 percent of that oil to places like America, the Emirates, and China. All that sweet, sweet black gold made The House of Saud one of the wealthiest families in the world.

Oil, though, is a finite resource. If Saudi Arabia stopped all oil exports today, assuming they don't find new reserves anywhere on their territory, they'd run out of oil in a little over 200 years. Of course, neither of these things is likely to happen. So we're looking at a time over the next century when the country will likely reach peak oil. That doesn't even consider changing attitudes towards oil versus sustainable renewable resources.

With the writing on the proverbial wall, the Kingdom is adamant about divesting from oil. But that isn't stopping them from going balls-to-the-wall now with production and distribution. To be fair, much of the pressure on the country isn't coming from its own population. With the ongoing war in Ukraine, and other geopolitical tensions, the United States and EU countries are trying to force Saudi Arabia's hand. President Biden made a special trip to the region to persuade OPEC leaders to make bigger commitments, which they did not do. The Crown Prince himself even went on record saying Saudi Arabia '… has announced an increase in its production capacity level to 13 million [barrels per day], after which the kingdom will not have any additional capacity to increase production.'[8] Sorry, Yankie, go find your medicine somewhere else.

It seems that little YouTube video I watched left a few key points out, huh? Regardless, it could have gone on for hours and still would have never given the truth. The Line is one man's vanity project cloaking itself within the guise of sustainability to give it supposed credibility, clout, and public approval. That's pretty much the textbook definition of greenwashing. What's worse is that it's the kind of greenwashing we'd expect to see from an oil company in the 1960s. If you're going to greenwash, at least try to do it well.

And while this may be a very egregious example of highly produced greenwashing, Saudi Arabia is not alone. Around the world, nations are engaging in record levels of state-sponsored greenwashing. We'll get into examples of these shortly. What I want to do first is explore the big question: why? Why would countries start adopting the same unsavoury practices corporations have been engaging in for decades? If governments (at least in perfect democracies) are supposed to protect their citizens, why throw all this away for vanity projects that will never stand up to scrutiny? The answer, dear reader, is that nations have just as much to misdirect us from as those big, bad companies.

In the last chapter, we talked about the Paris Agreement and how it was unique versus its predecessors. Remember that the most significant difference was in its approach to getting countries to act. Instead of legal mandates, as was done with the Kyoto Protocol, the Paris Agreement allowed individual nations to determine their own levels of support. The Pollyanna idea behind all this was to make fixing climate change equitable. Developed countries with money would feel obligated to invest. Smaller countries, maybe at the most considerable risk of feeling the effects of climate change, wouldn't feel as if they had to do as much.

Except this is the real world, and none of that worked out to plan.

Instead of making things more equitable, countries have used the notion of national contributions as an excuse to not act. I mean, *at all*. The Climate Action Tracker is an easy-to-use map that measures how compatible countries are with the Paris Agreement. They've monitored this since at least 2017, updating their charts regularly. In the Tracker's December 2022 map, there are exactly zero countries on track to meet their goals against the Paris Agreement.[9] All 190-something signatories are behind to varying degrees. Some, like Costa Rica, Nepal, and Norway are 'almost sufficient'. There are 'insufficient' countries like Australia, Brazil, Japan, and the United States. 'Highly insufficient' countries include Argentina, Canada, China, Egypt, New Zealand, Saudi Arabia, and the UAE. Then we get to the worst of the bunch: the 'critically insufficient'. In alphabetical order they are: Iran; Mexico; Russia; Singapore; Thailand; Turkiye; and Vietnam.

So now we can understand why some of these countries would go so far to distract us from what's really happening. Honestly, though, they don't even have to do too much. A simple example of this came up at COP-27.

I know I've already bored you enough with the history of COP, so I'll make this short. During the Sharm Conference, delegations erected these grand pavilions to showcase all the great things each country was supposed to be doing. The Aussies served delicious lattes and discussed the future of sustainable agriculture. Singapore's pavilion was airy and green, a nod to the city-state's focus on building in line with the natural environment. Under the theme, Lifestyle for Environment, the Indian pavilion was highly designed and futuristic, reflecting their development ambitions. None of these, as far as I can tell, mentioned anything about their ranking on the Climate Action Tracker, oil addiction, or subsidies to environmentally disastrous mining companies.

They also forgot to mention some of the people they're appointing to positions of power are the same ones causing the problems. As I write this, the UAE is readying itself to host COP-28. One of their biggest announcements to date has been the appointment of Sultan al-Jaber as chair of the Conference. This man will be leading all negotiations. The not-so-funny thing is he's also head of the state-run Abu Dhabi National Oil Company. Yep, they put an oil chief in charge of COP. It's basically a big 'fuck you' to everyone watching and testament to how much these governments don't care.

There are plenty of other examples, too. Former ExxonMobil CEO Rex Tillerson served as Secretary of State under Donald Trump. During the short-lived Liz Truss era, notable climate sceptic Jacob Rees-Mogg led energy policy. As Brazilian President, Jair Bolsonaro allowed for record levels of Amazon deforestation. All of this is just a game to them.

Reading some of these names, one might assume this is a left versus right thing. That's not necessarily the case. While it's true there are far more climate change sceptics on the conservative end of the political spectrum, much of the issue is that harmful policies are baked into the system itself. That's because the needs of business dominate how politicians act. Therefore, it doesn't matter who's in charge of legislation if they continue to answer to greedy corporate bosses. I don't say this to scare you but to open your eyes and demonstrate the scale of the battle we face.

This battle, though, isn't being fought on a level field. Much like business, there are nations doing far more to greenwash than others. As you can imagine, the biggest perps work hardest to distract. We've already seen the amount of time, money, and energy Saudi Arabia is investing to save face. All that is just smoke, mirrors, and a good PR agent. If we put NEOM and The Line through the filter of our three greenwashing categories, it's a whole lot of misdirection and green speak. This holds for most of the petro-states in the same way it does with Un-Sustainable companies. Any effort they put forth is going to be easy to see through.

What about the more insidious examples? You'd be surprised by how well some of these countries hide behind long-held clean-and-green personas. While we could fill books with greenwashed national projects, I'll spare you the pain. Instead, I want to look at three of the most glaring examples where image doesn't line up with reality: Australia; Singapore; and the European Union.

More Than Meets the Eye

If I ask you to think about a typical Australian scene, a few things will automatically come to mind. The first is world-famous beaches like Bondi, Wineglass Bay, and the Whitsundays. Then you've got kangaroos bounding effortlessly across open grasslands or Koalas hanging gingerly from eucalyptus trees. Probably the most famous image of all is the Opera House sitting like big white sails atop the reflective waters of Sydney Harbour. I'll let you in on a little secret. There's a lot more to Australia's landscape than what you see on picture-perfect postcards.

Queensland's Galilee Basin is about 300 kilometres inland from the gorgeous Whitsunday Islands. This massive land area is one of the world's largest untapped coal reserves. Excuse me—it *was* one of the largest. Over the past decade, India's Adani Group successfully lobbied the Australian state and federal governments to open the highly controversial Carmichael coal mine. In its original form, the mine would have been Australia's largest ever, with a capacity to pull 60 million tons of coal out of the ground every year for ninety years. Getting all that coal offshore to places like India would mean increased shipping through channels near the Great Barrier Reef. The mine would also stress scarce natural resources like water and displace Indigenous communities. This is in addition to the billions of tons of Scope 1, 2, and 3 pollution from the mine. Of course, Adani refuted all this with bogus stats (and enticing contracts).

Aussies weren't having any of it. In one of history's greatest upsets, the grassroots group #stopadani has successfully fought back against the mine. Their work against Adani and any company stupid enough to support the Carmichael project is a case study in best-practice activism. In major capital cities, vocal protesters would stand outside financial institutions, construction companies, and even Adani's offices to have their voices heard. Support bled into political platforms, media campaigns, and households around the country.

In my neighbourhood, thousands of kilometres away in Melbourne, there are at least a half dozen #stopadani posters in windows still today.

All their hard work paid off. The group put tremendous pressure on companies financing the project. As a result, much of the Carmichael mine's funding dried up. Institutions like '... Goldman Sachs, HSBC, JPMorgan Chase and more have ruled out financing coal projects with specific references to Adani Mining's activities.'[10] Going further, Allianz '... no longer offers single-site/stand-alone insurance coverages related to the construction and/or operation of lignite/coal-fired power plants and mines where lignite/coal is extracted.'[11] Adani was left scrambling for close to a decade. While Carmichael still began operations in 2021, it was only at a fraction of its intended scale. Moreover, it remains the only mine in operation in the Galilee Basin, even though the Queensland Government approved almost a dozen similar mines in 2010. That's a massive win at keeping all this polluting stuff in the ground where it belongs.

Even so, Carmichael isn't the only mine pockmarking large parts of the country. From Queensland to Western Australia, and most of the desert in between, global firms are creating an otherworldly moonscape (hellscape?). Over 350 mines are operating to pull out coal, minerals, and precious gems from the ground.[12] Overall, mining accounts for about 10 percent of total GDP, making Australia the fourth largest mining country in the world.[13]

The industry's power and influence are so vast they can get away with murder. In 2020, mining giant Rio Tinto got the green light to destroy a sacred Indigenous site dating back to the last Ice Age. In blasting the site to smithereens, they also permanently removed historical evidence of local peoples' continuous inhabitation of the area for the past 46,000 years. Initial uproar died down, and now mining companies continue unabated. The situation has gotten so bad the West Australian First Nations peoples formally petitioned the United Nations Committee on the Elimination of Racial Discrimination for intervention.[14]

The work of #stopadani has reverberated throughout global mining, creating a broader conversation on the industry's environmental impacts and pressure to stop operations altogether.

> The Stop Adani campaign is much bigger than a single mine—it has catalysed urgent conversations about the future of thermal coal in public forums and private board rooms. Banks and investors across the world are reducing their exposure to thermal coal as a result of this campaign.[15]

Yet it doesn't seem the folks in Canberra have received the message. Because these mines are outside major population centres (thank god), and because they are just so lucrative, Australia's political leaders think this a case of 'out of sight, out of mind'. In the halls of Parliament, voices of dissent get met with disdain and ridicule. The most famous response to calls for tighter environmental regulations has since become a popular meme. In 2017, future Prime Minister (then-Treasurer and eternal bozo) Scott Morrison brought a big lump of coal into Parliament. With a shit-eating grin, he told his colleagues and the country not to be scared of coal. It's what'll keep the lights on. In the words of British academic Paul Ekins, Australians '... always deliver the most shocking news in such an entertaining way.'[16]

As asinine as the whole scene was, and as much as it pains me to type this, Morrison was right.

Australia likes to call itself the lucky country, not because of its beautiful natural environment but because it's been able to weather global storms of recession better than most nations. People here are wealthier, on average, than any other country on Earth. Reflecting this is the ridiculously high cost of living and the massive spending power of Australian consumers. All this is thanks to one of the main pillars propping up the economy: digging things out of the ground. Yet given how on the nose this might be to some people, Australia is happily hiding this fact from the rest of the world.

This deception has led to long-standing accusations of state-sponsored greenwashing. In her excoriating 2022 piece for *The*

Saturday Paper, climate scientist Polly Hemming pulls no punches ripping Australia a new one. She notes Australia is unique in its levels of Government aid to polluting corporations, particularly within oil, gas, and mining, through unique certification and accounting mechanisms: '[I]n Australia the government is not only in on this game, but it is also running it.'[17] Here, it's not uncommon to see 'climate-neutral' oil and gas firms (for clarity, there is no such thing). Generous, dodgy offsets mean energy companies can claim they are pushing the boundaries of a sustainable organisation. As we've seen, political malfeasance also enables polluters to do what they like, where they like.[18]

The recently elected Labour Government makes this even more acute. They're waving the green flag of sustainability while greenlighting hundreds of new fossil fuel projects. For an outsider looking in, it also appears they're actively working against environmental protection. In 2022, Australia's Environment Minister Tanya Plibersek petitioned UNESCO to take the Great Barrier Reef off its list of endangered sites. The argument: we aren't as bad as the former (conservative) government, so you can trust we'll handle things correctly. You don't need to call us out. Given four mass bleaching events in seven years, and 90 percent of the Reef's coral currently bleached, it seems things aren't exactly under control.

Ultimately, Hemming posits, '... [i]f the world is to succeed in its goal of reducing greenhouse gas emissions, Australia must fail in its goal of increasing fossil fuel production and exports.' We can only do this when we look beyond the perfect façade of this so-called lucky country.

Trouble in Garden City

When it comes to the perfect country, Singapore is about as close as it gets. Since independence in 1959, the small city-state has grown from a land devoid of natural resources to one of the world's economic powerhouses. Labelled an Asian Tiger, Singapore has the world's second-highest purchasing power parity, the busiest container port, and is the regional home for most multinational corporations. Beyond economics, Singapore consistently ranks high on the United Nations Human Development Index. When one digs into the specifics, it's easy to see why.

You could fill pages with Singapore's accomplishments for its citizens. Given it has the world's fourth least corrupt public sector[19] that makes sense. Its advanced healthcare system contributes to Singapore having the world's fifth highest life expectancy[20] and one of the lowest under-five mortality rates.[21] Innovative policies in public housing mean nearly 90 percent of Singaporeans own their own home.[22] Internet speeds are the fastest in the world.[23] That's a good thing because when it comes to education, 60 percent have an advanced certificate or tertiary degree.[24]

It would be disingenuous to call this a façade. But Singapore's growth hasn't exactly been organic either. Like its lush gardens, the country has carefully manicured and manufactured this image over the past seventy years. The chief architect of this was Singapore's long-time Prime Minister, Lee Kuan Yew. As with any country, much of his work involved projecting a national image of strength. Surrounded by what were considered hostile forces, namely Malaysia and Indonesia, Lee created the Singapore Armed Forces. All able-bodied men over age eighteen are required to enlist, which is still a rite of passage to this day.

Unlike its resource-rich neighbours, Singapore has also had to find a way to differentiate itself economically. Initially, Lee thought of turning Singapore into the world's factory. While this worked to an extent, it wasn't until he prioritised attracting foreign talent and investment that the economy really took off. This, in turn, created a need for infrastructure and development that set off a virtuous cycle of growth. Once the infrastructure, economy, and society were stable, Lee looked to financial institutions. Attractive incentives have turned

Singapore into one of the world's biggest tax havens. Companies continue to flock to Singapore, most recently in a wave from Hong Kong.

Soft diplomacy was also central to the strategy. Singapore prides itself on being the world's 'garden city'. From Marina Bay, to its parks, to inland jungles, Singapore is definitely living up to its moniker. In fact, nearly 50 percent of its area is greenspace.[25] Not content with being considered one of the world's greenest cities, the Government has plans to take the top spot by 2030 through its Singapore Green Plan.[26] The Plan specifies a few really cool initiatives. By 2030, every Singaporean will be able to reach a park within a ten-minute walk. There will also be a 50 percent increase in green space and one million more trees.

Because Singapore is so central to global business, over my decade-plus in the region I've done quite a bit of work there. For those who haven't been, it's pretty tricky describing just how impressive a city Singapore is. It begins the moment you start your descent over the Malacca Straits, where out the aeroplane window you can see the hundreds of ships carefully aligned and patiently waiting for their chance to berth. Landing at Changi—consistently voted the world's best airport for at least the past decade—you can shop, eat, and even visit a butterfly garden, all before customs. Once through, you might want to cross the terminal and visit The Jewel. Housing the world's largest indoor waterfall, this mega-mall is seven storeys of even more shopping, dining, and sensory delights. Then, it's a short fifteen-minute drive into the city centre. Along the way, you pass through rainforest canopied highways, the gorgeous Marina Bay Sands, and exquisite Gardens by the Bay. Unpack your bags at any number of modern hotels, and the city is yours to explore.

That's why I was a bit taken aback when looking at the latest Climate Action Tracker. There, they have Singapore listed under the 'critically insufficient' column, the worst of the worst, when it comes to meeting their Paris Agreement targets. For a place that puts such an emphasis on clean and green, how could this be? Was there a flaw in the methodology, someone dozing off while crunching numbers, or was something more nefarious at play?

As with much state-sponsored greenwashing, it all boils down to the economic drive for development. What took the industrialised world almost 200 years to achieve, Singapore has been able to do in under half that time. Obviously, drive and vision are only two contributing factors here. Filling in vast swaths of the bay, erecting enough units to house an entire population, and building the infrastructure to make Lee Kuan Yew's dream a reality is a resource-intensive exercise.

Remember, all Paris Agreement commitments are up to the country itself. So the Climate Action Tracker is looking at how far Singapore is falling short of expected commitments versus its global peers. Given just how advanced the country is, this expectation is going to be high. In their analysis, the Tracker notes, 'Singapore's climate policies and commitments reflect minimal to no action and are not at all consistent with the Paris Agreement's 1.5°C temperature limit.'[27] If all countries followed Singapore's lead, the Earth would be on track for a 4-degree rise in global temperatures this century.

This analysis is down to a very simple key factor: Singapore's energy is dominated by fossil gas. The authors note just how difficult it will be for Singapore to make a clean-energy transition, especially considering its dependence on fossil fuels. Singapore, too, understands this and says most of its targets are 'dependent on international cooperation and technological development'. These goals are unrealistic as an export-oriented market, and with the rest of the world just as uncommitted to climate adaptation. As an energy source, fossil fuels then also underpin every other part of Singapore's commitments. That means that even with the most ambitious mitigation efforts, '... the effect of these policies in emissions is limited and does not compensate for the overall increase in energy demand.'

Interestingly, the Tracker also puts forward what a strong commitment would look like for Singapore, given its place on the global stage. They call this a 'fair-share target'. Unfortunately, the country is left lacking here as well. None of Singapore's plans aligns with what other advanced economies have committed. Singapore's commitments would lead to emissions seven times higher than what would be considered fair.

I take from all this that Singapore might not necessarily be engaging in state-sponsored greenwashing, at least not to the same levels as Saudi Arabia or Australia. Just look at all of the initiatives put forward to green the city beyond how green it is currently. Energy efficient buildings, infrastructure designed to blend and work within nature, and even conservation, all point to a country doing more than its fair share. Where this all falls apart, though, is in what it's going to take to get there. Making Singapore greener will require even more resources and non-renewable energy. It will cost us upfront to benefit down the line. How we view this through the lens of greenwashing all comes down to whether or not we're willing to trust Singapore with this investment.

In speaking with Evelyn Hussain, an expert in human-centred circular design, she believes trust in Singapore is well placed:

> While the Government might be putting forward ambitions at a global level, it's just as important to examine what's happening locally. As a Singaporean, I've seen just how engrained sustainable thinking is across all parts of our society. Yes, there are underlying factors that may be working against us. But if history is any indicator, it's unwise to think of Singapore as an underdog.

I, for one, am happy to have the people of Singapore on the planet's side.

The Changing Face of State-Sponsored Greenwashing

Nearly a hundred years after World War Two, the fear and reality of populism still have a stranglehold on Europe. While the radical right-wing neo-Nazi fascists might get the most airtime, what's most insidious are the parties, a slight tick to the left, that have infiltrated policy-making. They've done this by recognising that changing attitudes to critical issues, like climate change, necessitate changing approaches to messaging. Obviously, the electorate isn't going to laugh anymore if you bring coal into Parliament. Neither will they stand for policies and practices that literally kill their families. Thus, political parties on the right are now taking a more nuanced approach to keep with their time-honoured positions while presenting a fresher, more supportable face. The frontline of this fight is happening right now across the European Union.

Unlike right-wing climate rhetoric in the United States, European conservatives don't deny the existence of climate change. Instead, they question how much humanity is responsible and what we're supposed to do. That's a very different proposition and one more likely to garner attention since it's not nearly so off the wall as a MAGA populist. Marine Le Pen's *Rassemblement National* party takes what they call a 'humanist climate policy'. This mantra means people come first, which sounds good until you realise the planet comes much further down the policy docket. There's also a focus on immigration, one of Europe's most significant policy issues over the last decade. Unlike what makes logical sense, their party line isn't about fixing climate issues to reduce mass migration from places like Africa. Instead, it's focused entirely on migration and ignores its underlying causes. One can't deny climate change is a source of many of our current problems. Yet, these populist parties are working hard to avoid having the planet entering into any conversation.

There's also going to be a reckoning in the not-too-distant future regarding funding from Brussels. A member-state far-right party won't curtail the EU's climate ambitions. Instead, Brussels will likely tie funding to climate pledges, actions, and achievements. Agreeing to address climate change becomes very appealing when you see that lucrative carrot dangling in front of you. For this reason, European countries may '... end up with a populist government that's going to be terrible on very basic rule of law and human rights issues, but quite good on climate policy because the deal will have been so sweet, in fact, impossible to turn down.'[28] That will give them further cover, and widely accepted talking points, when they get up in front of the world.

If we're being really honest, this double-speak isn't a new thing for Europe. For decades, EU nations have portrayed themselves as the epitome of sustainable, green, eco-friendly societies. On the surface, this seems very accurate. Look at Scandinavia, with its pristine waters, clear skies, and healthy people. There's Germany, where trash sorting gets almost religious sanctity. The Dutch invented wind farming. French viticulture prides itself on ecologically sound practices. Italian nonnas make next-level ragu using only locally sourced ingredients.

While these images do have some credence, they belie deep-seated systemic issues political leaders across the EU would rather you not know. Much like Singapore, we must look beyond policy documents at what drives and fuels growth. Until recently, nobody really paid attention. Then a certain Mr Putin threw a big old oily spanner in the works when he decided to invade Ukraine. Russia was Europe's gas daddy. In 2021, prior to the invasion, they were supplying 40 percent of the EU's natural gas.[29] Russia had Europe over a barrel, too, importing 30 percent of the continent's oil.[30]

The sudden global attention on Europe caught the entire EU with its pants down. Now they had to do something. Germany, the largest Russian gas importer, was first out the gate. The country's Finance Minister, Christian Lindner, said, 'Germany had completely diversified its energy infrastructure since Russia's invasion of Ukraine.'[31] The EU also makes mention of Russian gas imports now being close to 15 percent, less than half of what they were before the invasion.

What the German Minister and EU representatives aren't saying is what's filling the gap? No one's stopped importing fossil fuels. Instead, Germany and the rest of the EU now get their gas and oil from places like Australia, the US, and North Africa. In a classic case of selective disclosure, the public only gets half the story. It's all about 'we've gotten rid of Russian gas and oil', but not so much about 'we've replaced it with gas and oil from somewhere else'.

Across the EU, plenty of above-the-fold versus below-the-fold announcements are going on. The Italian Government made a big splash shelling out billions of euros, the most in the Union, to subsidise converting properties into low-carbon eco-homes. This investment is overwhelming the building industry, a sector known to be highly polluting, along with available labour and materials. The Netherlands prides itself on an outstanding work-life balance. At thirty hours, Dutch citizens have the developed world's shortest work week.[32] This landscape has attracted talent from around the world, meaning the Netherlands has basically run out of space for more people. With only 40 percent of employees working full time, there's also not enough of a workforce to service the economy. Three-hour lines to enter Schiphol Airport last summer demonstrate this fact. Sweden has long been considered the world's most sustainable country. Just don't talk about its forestry industry which relies on clear-cutting all but 3 percent of the country's forests.[33] This destroys biodiversity and releases enormous amounts of carbon during logging. While the EU urges against clear-cutting, Sweden has told Brussels it has no business dictating the country's forestry strategies.

None of this is without historical precedent. Colonial-era Europe benefited markedly from exploiting populations thousands of kilometres from the continent. From the Transatlantic slave trade in Africa to spice wars as far away as Indonesia, no one thought to ask why as long as people were getting richer. As unsavoury as it may sound, how different is any of this today? It's no secret climate change has a far more negative impact on communities of colour, especially in the developing world. In many respects, this is just a case of Europe being Europe.

But now the proverbial chickens have come home to roost. Flooding across Germany, Belgium, and Austria in 2021 caused hundreds of deaths and billions in damages. Europe, notably the United Kingdom, suffered a historically intense heatwave during the summer of 2022. While tabloid newspaper *The Sun* ran bikini-clad cover photos about how fun all this was, temperatures reached a scorching 40 degrees Celsius (104 degrees Fahrenheit). Dramatic scenes on social media showed commuter trains in Italy and Spain making their way through the middle of massive forest fires. Drought followed, with nearly every country in Europe impacted. In France, villages ran out of drinking water. There was so little water in the Rhine that shipping stopped. Spain rationed water supplies.

Whether these Biblical events will cause a come-to-Jesus moment is anyone's guess.

National development isn't a new thing.

National development isn't a bad thing.

National development isn't something that's going to stop anytime soon.

Like the countries in this chapter, greenwashing at a national level is far more nuanced a subject than what we would think on the surface. It's not as easy as saying, 'development bad; green good'. Nor is it necessarily about placing blame on some and giving a pass to others. While I may have highlighted Australia, Singapore, and Europe, I could have easily picked from any place on Earth. That's because no one is innocent, and there are plenty of examples to choose.

The key difference here, and what really makes this true state-sponsored greenwashing, is in the image each nation portrays. The world knows China has high carbon emissions. The world knows the Middle East traffics in oil. The world knows India's streets are choked with smog as the country builds for the future. But do they know about

Australia's obsession with coal? Do they know how much Singapore is gobbling up our planet's finite resources? Do they know how Europe is engaging in shifty, post-colonial environmental offsetting?

No.

Which is why it was so important for me to pull out some of these cases for this book.

Anyone who has been through addiction recovery knows the first step is always recognising the problem. The unfortunate part is that, unless you're reading this from the halls of power, it's not really clear what you can do to effect change. People power only goes so far, especially when we're up against what I'm pretty sure my mom would have called 'the man'. For those sitting in parliament, congress, or the duma right now, though, let's chat.

You've got a problem. This is your intervention.

I say this from a place of love. Your list of addictions is long, even if we narrow it down to the tiny slice of the pie that negatively impacts the planet. The big addiction, obviously, is to *oil*. Lobbyists and corporate donors have you eating out of the palm of their hands. Complicating this is your weird marriage to *old ways of thinking*. Oil is great, you think, because it got us from there to here. But it's 2023, and many other resources are at our disposal. These resources could easily replace oil, do right by the planet, and even benefit you if played right. Yet you do nothing because of *complacency*, your third big addiction. Doing what you've always done is much easier when you only consider the short-term. When the focus is on winning the next election, I guess there's no room for long-term thinking.

Coming to terms with having a problem is just the first step of many. You've got to then find support to keep on track with your recovery. That support will help you detox and find new routines to occupy your mind. It's hard to shoot up when a bunch of people enthusiastically remind you how good life can be without that junk. In the same way, tapping all those finite resources will be difficult if there are more logistically, financially, and environmentally friendly options to power the world. The key is to stop being complacent, get off your ass, and act.

But it's not just your addictions that pose a problem. Like an annoying mosquito, *we're* your problem now, too. For a long time, you 'leaders' have assumed we're just simpletons easily distracted by shiny objects. As long as you kept us looking in one direction, there's no way we would ever find out you've been lying to our faces. Hell, you even brazenly parade your misdeeds, almost daring us to do something.

Well, we've taken you up on your offer. Now, we're getting to work.

Which leads to the biggest step of all: prioritising what's important. It's clear the environment isn't important to you. There's no question you don't actually care about the needs of your constituents, either. So, then, go ahead and be selfish! I'm permitting you to prioritise number one. Think about all the ways divesting from your addictions—oil, outdated thinking, complacency—can materially benefit you. We'll reach peak oil sooner than you think. How will you get those millions in contract revenue if you wait until the last minute to act? Switched-on businesses are innovating faster than ever. Do you really think they care about your laughable position on climate change? If you think things are difficult now, consider how much more difficult they'll get over the next decade. In the immortal words of Charles Darwin— 'adapt or die'.

Chapter 13

Hyper-National Organisations

Whether you agree with it or not, globalisation has dramatically changed how the world works. Not so long ago, you were only accountable to the local authorities. Depending on when and where you lived, that could be a shaman, sheriff, or preacher. Then things started to expand, and national governments began to take the lead. Following two destructive world wars, humanity came together and bumped all this up a notch. The creation of organisations like the United Nations, World Bank, and International Monetary Fund ensured stability in an increasingly globalised world. Entities like the European Union allowed for easier movement of goods and people across borders. Everything was working swimmingly.

Today, though, things have gotten out of hand. We've gone beyond these multinational organisations and now see dominance from hyper-national organisations. These groups don't necessarily operate within the bounds of everyday politics. Most of them aren't technically political entities at all. Yet they influence policy, shape our ways of thinking, and have more influence on our day-to-day lives than most people realise. I'm talking about sporting leagues, think tanks, and donor communities.

An odd grouping, you might be saying to yourself. I agree. But each is guilty of wielding far more power and influence on today's society than they did only a decade or two ago. With that power comes an even greater need for transparency and accountability. Nowhere is this more pressing than how these organisations present themselves as stewards of a more sustainable future. Like corporations, they love to talk about all their eco-friendly, sustainable initiatives. Some make grand sweeping claims, like using the Olympic Spirit to inspire environmental consciousness. Others drone on about 'selflessly' spending millions on this or that worthwhile project. But does it? And are they?

In this section, I want to examine how greenwashing permeates even these do-good, hyper-national organisations. From oil sponsorships at the World Cup to blood on the hands of the elites in Davos, let's see whether actions match all the statements we keep hearing.

Greening the Pitch

From the moment Sepp Blatter opened the envelope awarding Qatar hosting rights for the 2022 FIFA World Cup, the desert monarchy was in for an uphill battle. Almost immediately, allegations of corruption began to swirl. Reports confirmed that Qatari officials bought votes from FIFA committee members in the order of millions of dollars. As part of a broader investigation into corruption in the world soccer body, half of those who voted in the 2022 bid process have since been fined, banned, or jailed for related fraud.

Then there was the pesky issue of Qatar's human rights record. Like its neighbour, Saudi Arabia, let's say it wasn't that stellar. From the vaguest semblance of women's rights to execution for those in the

LGBTQIA+ community, Qatar was operating more like a backward Medieval fiefdom than the modern country it was trying to portray. Issues compounded in the years leading up to the big event. The exploitative *kafala* system essentially brought in labour from across the region to work in slave-like conditions, building the Cup's infrastructure. By some estimates, at least 6500 workers died.[1]

Facing all this, Qatari officials must have met in an air-conditioned boardroom high above the desert sands, all dressed in their best thobes (unsurprisingly, the board was made up entirely of men). They could fix their human rights practices, but that'd be too difficult. How about reforming the *kafala* system? Maybe a bit, but not too much since it works so well. Stop bribing officials? What were they supposed to do with all the oil money spilling from their wallets?

How about going green?

Yes! What a perfect distraction from all the global hubbub around human and labour rights. Of course, nobody in the room believed they'd do anything to change. They were going to publicly wrap themselves in the comforting blanket of environmental sustainability while privately sticking to what they knew best. Who would call them out on this, especially with the best public relations machine money could buy?

That machine quickly went to work. FIFA and Qatar announced this would be the world's first carbon-neutral Cup. A big part of that would be thanks to the most advanced, greenest stadiums ever built, including one made entirely of reused shipping containers. They even had the numbers to prove it!

Except, none of that was true. Through a series of rather ingenious calculations, which relied heavily on carbon offsets and projects that didn't even exist, organisers seriously fudged the numbers.

The tournament's organizers calculated the total emissions associated with the construction of all the stadiums, then divided that figure by the average lifespan of each stadium, estimated to be approximately 60 years. Because the World

Cup lasts for only one month, organizers then concluded that the tournament was responsible for only one month's worth of emissions spread across that 60-year estimated span.[2]

There's also upkeep for those stadiums, with pitches that need 50,000 litres of desalinated water every single day,[3] as well as all the flights and other infrastructure associated with the Cup. Although final numbers aren't in yet, many environmental scientists say FIFA and Qatar underestimated their carbon emissions by at least 1.4 megatons.[4] In total, some have argued this makes the 2022 Qatar World Cup the single most polluting event (aside from war) ever put on by human beings.

Like The Line, the Qatar World Cup was a blatantly juvenile attempt at greenwashing. Nobody was buying what the Qataris were selling (do we see a trend here?). The thing is, FIFA and Qatar aren't alone. These mass sporting events are notoriously bad for the environment and local economies. Yet they keep touting themselves as leading the way in sustainability. Of course, the big daddy of them all is the Olympics. Yet even though there are contractual agreements in place with host cities on environmental sustainability, organisers easily overlook the fine print. An in-depth analysis in the journal *Nature Sustainability* found each subsequent Games has been less sustainable than the last.[5] Three of the most recent Games—Tokyo, Rio, and Sochi—were the worst of all.

Study after study derides the winter and summer Olympics as more trouble than they're worth. It's gotten so bad that the International Olympic Committee is having difficulty finding cities that want to host. The Games have become the poisoned chalice nobody wants to touch. Nobody, that is, except corporations. Around the world, corporations are throwing a lot of time and money at big-name sponsorships with organisations like FIFA and the IOC. Like Qatar, they hope to improve their reputations by using sport to greenwash.

While it may be difficult to see the link between sport and greenwashing, there are very particular reasons corporations invest in this strategy. Firstly, few marketing channels have as big a platform as sport. There's a ready audience already engaged. When it comes to bang for the buck, you don't get much better than that. Companies can use soft power to associate themselves less with destructive behaviours or practices and more with sports entertainment, fun, and camaraderie. This translates to a stronger social license to operate, which is how much a particular community accepts a business or organisation. You see this at a local level when some mom-and-pop shop sponsors a school soccer team. The same principle gets pumped up on steroids the bigger the audience gets.

Beyond these massive quadrennial events, platforms for sport greenwashing are seemingly endless. To see just how far they extend, let's look at the most sport-hungry nation on Earth: Australia. When I say Australians love their sport, I'm not exaggerating. Nearly 90 percent of Australian adults participate in some form of organised sport or physical activity.[6] Almost 80 percent call themselves sports fans.[7] In the state of Victoria, home to Melbourne, there are even two public holidays for sporting events. That's because sport is big business, contributing A$50 billion to the economy yearly.[8] It also means sport is a big opportunity for corporations to market to millions.

Research conducted by the Australian Conservation Foundation[9] looked at corporate sponsorships in Australian sport. Their report, *Out of Bounds*, found nearly 1500 major sponsorships across a wide sampling of divisions, activities, and levels. Of these, the vast majority came from retail operations and consumer goods. Industrial, construction, and transport comprised 9 percent of the total cohort. Auto contributed a further 7 percent. What the researchers were mainly looking for, though, was the level of involvement from Australia's most notorious industries: oil; gas; and mining.

The report calls out the uniqueness of coal, oil, and gas sponsorships. While they make up less than 4 percent of total sport sponsorships by number, the dollar amount invested sets them apart. We're not just talking about having a sign at a rugby match. These companies are paying for naming rights on stadiums. You've also got mining giants BHP, BP, and Santos all slapping their logos on Aussie Rules football teams. Ampol is a key sponsor of the National Rugby League and the naming rights sponsor of Australia's most-watched sporting event, State of Origin.

These figures are just as striking when we pan out to a global level. A 2021 survey by the New Weather Institute[10] explored the prevalence of corporate greenwashing in sport. In a random sampling, they found 258 sport sponsorship deals with companies profiting from high-carbon-emission products or services. The range inculcates every level of sport: from cricket to cycling to rugby; at the Olympic level down to community groups. Football (soccer for Americans) was the biggest beneficiary of corporate sponsorship, making up nearly a quarter of all deals surveyed. Some of those deals came from the most polluting companies on the planet, including Russian gas giant Gazprom. The firm made a US$90 million multi-year sponsorship deal with FIFA in 2018, paid US$22 million to extend an agreement with German club Schalke, and invests almost US$55 million a year sponsoring the European Football Championship. The silver lining is, since the publishing of this study, Gazprom has had several of these contracts cancelled given the Ukraine invasion. Unfortunately, other companies have just taken Gazprom's place.

But it's not only oil and gas taking part. Other highly polluting industries have spent many years and a lot of money using sport as a front. In 2021, the Australian Open—the first Grand Slam tennis competition of the year—came under fire from activists. Someone snapped a photo of an athlete mid-serve surrounded by the logos of three corporate sponsors: Santos; Kia; and Emirates. You couldn't ask for a better grouping of companies from industries most likely to use sport for greenwashing.

The New Weather Institute study found, by sector, the auto industry is the most prominent advertising sponsor of sport. Car companies spend 64 percent of their sponsorship budgets on sport alone, with Toyota being the largest sponsor overall. After automotive, the airline industry comes in second. Not surprisingly, Emirates was the leader of the pack in their sector. Of course, a lot of this sponsorship is simple marketing spend. They want their company to stay top-of-mind with global audiences. We have to remember, though, the residual impact of all this. Don't think the marketing masterminds at Toyota, Emirates, or any other company aren't considering all options to gain that valuable social license to operate.

Not to be outdone, sports organisations themselves are also dabbling in the greenwashing waters. We now see wider sporting bodies and individual teams publishing sustainability reports and environmental metrics. This is a definite move in the right direction! We must recognise that as a first go, there will always be missteps. I've got zero issues when these mistakes are used as valuable lessons and corrected. If comments by US National Basketball Association (NBA) commissioner Adam Silver are anything to go by, though, we might be in trouble. Speaking at the Aspen Institute's inaugural climate change conference (don't worry, we'll get to them, too), he claimed the goal of American sporting teams was to '… halve greenhouse gas emissions by 2030 and reach net zero by '40.'[11] He said this in the same breath as lauding the NBA for its position leading the transition to cryptocurrencies. Crypto is not only a bad financial investment but also terrible for the environment. The cognitive dissonance between sporting goals and Silver's comments is just one of many issues to address if US sport genuinely wants to make a positive impact.

To guide these organisations along their first steps in sustainability, the United Nations created the Sports for Climate Action Framework.[12] Over 300 sporting organisations worldwide, including the IOC, NBA, and Liverpool FC, signed the Framework almost immediately. Clearly, they saw this as a great way to get a leg up in helping the Earth, right? Eh … not so fast. The Framework didn't have any concrete targets, opting for broad language around reducing overall climate impact. Without teeth, the Framework became just another useless piece of paper. The UNFCCC surveyed signatories one year after the Framework's launch, asking how they were going measuring emissions. Over 60 percent hadn't even started yet.

It was all a way for these organisations to get excellent publicity and then fade into the background when it came time to do the work.

Given its massive platform, though, sport can have a tremendously positive impact. The first step on the road to stopping greenwashing is more robust regulation. It's not like this is without precedent. Tobacco was a major sponsor of world sport up through the 1980s. Australia's three largest tobacco companies were also the three most prominent sponsors of competitive sport in the country, investing A\$20 million in 1989 (that's A\$51 million today) just for Rugby League in New South Wales.[13] RJ Reynolds Tobacco, makers of Newport and Camel cigarettes, openly boasted about sponsoring 2736 separate sporting events in 1984.[14] With changes to advertising regulations, and public attitudes, good luck finding a cigarette ad anywhere near an event now. We're seeing the same happening today with alcohol sponsorships. In France, for example, soccer uniforms must be free of any alcohol branding.[15] That even applies to teams coming in from overseas.

Organisers also need to start looking at innovative alternatives to green sport itself. It doesn't seem truly worthwhile to pick a different Olympic host city every two years. Then there's the time, cost, and environmental impact of building the infrastructure associated with the Games. I've already mentioned the dwindling supply of willing candidates. So, why not just have a permanent home for the Olympics? Ideally, one city with all the infrastructure already in place can host again and again and again. There's also game scheduling in other sports that, with some clever adjusting, can eliminate a lot of unnecessary travel. Covid-19 necessitated some of this, particularly in the big business of US major league sports. Research has shown that if these adjustments and ways of working were made permanent, there would be a 22 percent reduction in air travel emissions annually.[16]

Lastly, the tide of public opinion is moving in favour of the environment. We already know global consumers are hyper-aware of sustainability issues, with millennials being the most switched on. Consumers want companies to stand for more than just profit, which extends beyond the supermarket shelf and into sporting organisations. I would argue it's even more acute given people's personal links with 'their' teams. Many are probably sitting in the stands, or watching on TV, wondering whether they want to associate themselves with polluting oil and gas companies. In Australia, over half of the population believes fossil fuel sponsorship of national teams must stop. Even more liken these to the cigarette ads of old.[17] It'll be interesting to watch, post-Qatar, whether all the corruption will actually lead to more people paying closer attention and changing their attitudes.

From Davos, With Love

I love *Veep*.

Honestly, it might be one of my favourite shows ever. Its rapid-fire, witty political banter and non-stop shade would have me in stitches. The HBO show follows Julia Louis-Dreyfus as Vice-President turned accidental President, Selena Meyer. Her rag-tag bunch of misfit advisors ran the gamut of political stereotypes: there was the ever-present shadow of the assistant; overly knowledgeable stats guy; bumbling drunk chief of staff; the press secretary who can't string a sentence together; and the staffer willing to do anything to get a leg up. They make political errors at every turn and use back-room dealing to enrich themselves while pissing off friends and allies. It's like watching the Trump administration, except a hell of a lot funnier and much less damaging to the global political order.

True to real life, building political capital happens as much on Capitol Hill as it does behind closed doors. During the show's final season, we see President Meyer testing the waters for an election bid. The cast finds themselves on the slopes of Aspen at an exclusive 'Discovery Weekend' hosted by a megalomaniacal recording billionaire, Felix Wade. A major party donor, anyone running for office views him and his gaggle of twink hangers-on as kingmakers. Saying as much would be too gauche, so the whole weekend is wrapped up in conversations about how to save the planet, yadda, yadda, yadda. At one point, there's even a half-day forum, 'Our Clean Coal Future', sponsored by a fictitious Coal Council. Perhaps this is all closer to reality than I realised.

Really, prospective candidates are trying to massage the ego of a man who's 'addicted to disruption' (although he's too idiotic to understand more than the talking point). He changes his mind repeatedly, pledging support for one candidate and then moving to another five minutes later. The key, they find out, is to repeat the last two sentences he says. That's the level of acumen this guy is bringing to the table. An excellent review of the episode notes that Felix '... completely captures a side of political maneuvering [sic] we don't really think about, clueless rich people who shower politicians in money so dark it would be shot entering its own apartment.'[18]

As I'm writing this, the yearly wank fest for the rich is happening in Davos. While technically an annual meeting of the World Economic Forum, Davos has come to represent the growing chasm between unrealistic elite aspirations and the reality most of us actually face. The parallels with Veep's Discovery Weekend can't be understated. Just look at the guest list: a group of elite, wealthy, powerful individuals.[19] Some of the big attendees this year included Al Gore, John Kerry, and fifty-one heads of state. Wall Street reps like Jamie Dimon and CEOs from Shell, Amazon, and Moderna were there, too. These people make more money in an hour than others make in a year. So when I say they're out of touch, I don't feel I'm exaggerating.

What exactly are these elites supposed to be doing? The stated mission of Davos is to demonstrate entrepreneurship in the global public interest. Looking through the 2023 themes, agenda topics, and speakers, everything seems to point towards supporting that mission. The overarching theme is exploring 'cooperation in a fragmented world'.[20] With geopolitical tensions and the rise of nationalism, this makes sense. Topically, sessions explored labour, cost of living, the global recession, extreme weather, and the ongoing war in Ukraine. On paper, it all checks out.

At this point, it's important to remember that greenwashing, boiled down, is simply a mismatch between words and actions. I'm inclined to point out that not all greenwashing is created equal. There is innocent, accidental greenwashing. There is also nefarious, purposeful greenwashing. I also firmly believe the bigger game you talk the more accountable you must be held. With all that in mind, has Davos accomplished what it has set out to do?

That's a tough question to answer when the mission is so damned vague. There was certainly a lot of talking going on, with over 200 sessions and thousands of panellists throughout the week, in addition to informal sideline meetings. Maybe *that's* where attendees demonstrated entrepreneurship—because it was rare to see it on stage. I mean, these are global leaders and captains of industry. They'd know better than most the importance of concrete deliverables. Beyond yammering, what the world's getting from Davos is as clear as mud.

What about motivation? This is as important a greenwashing indicator as delivering on KPIs. For some big-name attendees, you've got to question why they're there. Along with monied business leaders and influential politicians, others are more comfortable in Hollywood than in global economic meetings. This year some of the most recognisable names were will.i.am, Idris Elba, and Yo-Yo Ma. A special shout-out goes to Nas Daily, whose biggest claim to fame is recording 1000 one-minute-long videos for social media. Like COP, are people going because they are experts who can influence change? Or, as is likely the case, are they simply there for a few photos and a nice press release?

Then, of course, you have the usual suspects who need no introduction. We know exactly why these folks are hanging about. There's BHP and Fortescue, Rio Tinto, SINOPEC, and Chevron. Petronas, Shell, and Total. Our old friend Adani was there, too. All of them, plus hundreds more, are partnered with the World Economic Forum. As we discussed with sport greenwashing, companies that align themselves with global think tanks are also trying to build the all-important social licence to operate.

Useless conferences filled with endless panels and speakers, self-interested attendees, and dodgy sponsorship are nothing new. Even if they're talking about sustainability and still delivering nothing, this doesn't inherently mean it's all greenwashing. Davos, though, really talks up just how important it is to the future of humanity:

> The Annual Meeting will convene leaders from government, business, and civil society to address the state of the world and discuss priorities for the year ahead. It will provide a platform to engage in constructive, forward-looking dialogues and help find solutions through public-private cooperation.[21]

The people in those rooms have zero problems imagining themselves more intelligent than the rest of us, with all the answers and most of the resourcing (but none of the willingness) to get things done. In my world, that makes them the biggest greenwashers of all.

One guest of the 2023 Davos meeting noticed the same thing and used his platform to call out the hypocrisy. None other than Al Gore, politician, Nobel-prize winner, and noted environmentalist, was visibly seething while sitting on the main stage. When it was his turn to address the plenary, who had just heard from influential business leaders and politicians, he let rip: 'Every piece of pro-climate legislation … the oil and gas industry and the coal industry, they come in and fight it tooth and nail … to stop progress.'[22] He added that international institutions, like Davos itself, needed reform if they were ever going to positively influence change.

Oh wow, did they not like that. The World Economic Forum's own recording of the presentation is overlaid with almost comical music to tone down the very salient message. Headlines from major newspapers called Gore unhinged, fiery, and jokingly said he was having a meltdown. This, dear reader, is what we're up against. So much for speaking truth to power. It doesn't work if the most powerful forces in the world don't want it to work.

But Davos and the World Economic Forum aren't the only kids on the block. Countless other similar institutions are putting on meetings and conferences throughout the year. Groups like The Clinton Global Initiative, Milkmen Institute, and PopTech hold events with almost identical missions. Some gatherings are for specific industries. Media moguls like the Murdochs meet in Idaho each summer for the Sun Valley Conference, while bankers go to the Jackson Hole Symposium in Wyoming. Other events give a platform for the latest thinking, like the Aspen Idea's Festival or TED Conference. You could even count the conspiracy-laden Bohemian Grove meetings among this lot.

The segments they target are not materially different: ultra-rich power brokers looking for an excuse to meet. The topics are practically carbon copies, involving vague language around innovation for change, our shared digital future, or building tomorrow today. Hell, most are even neighbours. Rocky Mountain property values must be through the roof with all these highflyers darting in and out.

Overall, we must place what are ostensibly networking events for the wealthy and powerful in the proper context. Doing so is often tricky with all the media hype surrounding these events. Headlines tout their excellence, like Martin Wolf of the *Financial Times* opining about 'What I learnt in my days on the mountain in Davos'. Papers position these meetings as if they're the first of a kind. *The Aspen Times* notes a 'search for environmental identity' at the city's annual Idea's Festival, adding the not-so-groundbreaking finding 'a personal connection to nature helps drive climate solutions'.[23] They're even given a wow factor, like how NPR labelled the Sun Valley Conference a 'Summer Camp For Billionaires'. Talk about green sheen.

Davos and its peers aren't operating inside some vacuum. They are just one part of an entire ecosystem working on environmental and social issues. With all the media attention and big-name celebs swanning about, the world has forgotten this. We've grown to give these business meetings more credit than they're due. We think a bunch of elites meeting once a year will solve our problems and build a more sustainable future. Maybe it'll move the needle a bit. But much like the United Nations, I'm not sure these happenings are meant to advance anything. If their track record is any indication, they're just a giant exercise in assuaging guilt.

Yet, like sport, these organisations have the potential to make a genuinely positive impact. The biggest impediment to doing so is the gap between their reality and ours. I'm convinced the folks at Davos, Aspen, or Sun Valley still believe we need to persuade people there's a problem. That ship's sailed! It's no longer about telling people the world is on fire. Today the goal has to be action, action, action. Seeing another panel of white men in expensive suits lecturing the plebs isn't a vibe. I want to see them roll up their sleeves and get to work.

That's why there needs to be a broader dialogue across all parts of society. Davos pays lip service to this in their Open Forum, a free program available for all … who can afford to take time off work, fly to Davos, and stay for the week. More accessible events, like South by Southwest, still favour attendees and panellists from the developed, English-speaking world. The problems these attendees face are very different from people experiencing the impacts of climate change firsthand, translating into just another us-versus-them chasm. Alternative events like the World Social Forum attempt to counterbalance things. The giants easily suck up all the airtime, meaning traction is next to impossible for the new, more inclusive kids on the block.

But I'd venture to guess that many people with invitations 'lost in the mail' may have the answers attendees say they're searching for. The key now is to throw those doors wide open and let them have their say. Unfortunately, as Bloomberg's Allison Schrager points out, the bigwigs at Davos might not be so receptive to this idea. At a dinner full of global CEOs during the conference, the bulk of her conversations were attendees 'hating on Gen Z'.[24]

A Rose By Any Other Name

As we've seen, corporations have a desire to put their best face forward to the world. This desire is even more insatiable for those actively trying to create some *Mad Max* hellscape. They aren't content with the massive platforms sport offers. Nor are all the slots on all the panels at the highest-level events going to do the trick. Now, these greedy organisations are coming after some of our most prized and cherished institutions.

When we began our discussion of greenwashing hundreds of pages ago, there was a brief section on astroturfing, a practice where corporations covertly sponsor organisations or initiatives that seem to, on the surface, be doing good things. In reality, they're no more than a lobbying front. We called out oil giants as particularly fond of this type of greenwashing. News recently came out about Exxon's extremely accurate data around climate change. The findings show, 'ExxonMobil didn't just know "something" about global warming decades ago—they knew as much as academic and government scientists knew.'[25] So they covered things up, using these front groups to throw the public off the scent.

If you're thinking, *but that was the 1980s, John. People didn't know as much or have access to the information they do today*, then I'm sorry to say history is repeating itself.

You might remember Adani, the Indian mega-conglomerate that dabbles heavily in coal. They featured in our discussion on state-sponsored greenwashing in Australia. When they're not blasting big holes in the ground, Adani likes to see what exhibitions are showing at the world's finest art galleries. Their absolute favourite is the Science

Museum in London. How do I know that? Look at how in bed the two are with each other.

In a tweet from October 2021,[26] Adani chairman Gautam Adani proudly announced they were funding the new 'Adani Green Energy Gallery' at the British Science Museum.

Read that sentence again.

Once more.

The tweet continued. 'The new gallery will explore how we can power the future through low carbon technologies. It will be a reminder of the power of the sun and the wind in our daily lives.' A company deep in the coal and mining sector had the balls to position itself as a green energy leader. How do they square this with reality? Apparently, a new arm of the conglomerate aspires to become the world's largest renewable power generating company by 2030. The new gallery would be a way to legitimise these dreams.

The public backlash was swift. Within weeks of the announcement, a prominent Museum director and several board members resigned. One said staying on the board would give the '... false impression that scientists believe the current efforts of fossil fuel companies are sufficient to avoid disaster.'[27] Artists pulled their pieces. Activist groups, including Extinction Rebellion and the UK Student Climate Network, held days-long protests. Even 400 teachers from across the UK have signed an open letter boycotting taking their students to the Museum.[28]

The Science Museum isn't a stranger to criticism, and they were left unmoved. Perhaps it's because of some other prominent Museum patrons like BP and Shell, the latter of whom sponsors the exhibition, 'Our Future Planet'. A few rabble-rousers were not going to stop this cash cow. Defending their doubling down, the trustee chair showed her absolute naivety in saying, '... where a company is showing a willingness to change, our leadership team believes it is valid to continue to engage.'[29] Except, dear Dame Mary Archer, that's not what's going on here.

Adani's new gallery will open in October 2023.

Digging deeper into the world of corporate art patronage, it became clear the opportunities for greenwashing were far more pervasive than I realised. In the UK alone, BP sponsors such vaulted institutions as the British Museum, National Portrait Gallery, and Royal Opera House. Like greenwashing in sport, these are all ways of improving the company's social license to operate. None are as duplicitous as Adani or Shell talking about building a sustainable future. But they are greenwashing, nonetheless.

Which piqued my curiosity. If these companies were using the arts to launder their dirty practices, would they do the same by donating to other charitable causes? To be clear, there is absolutely nothing wrong with corporate donations to charity. I'd love to see more of it as long as solid governance mechanisms are in place. Problems arise, though, when the donor uses the opportunity to greenwash. Even worse is when they think a big cheque entitles them to dictate what the recipient can and cannot do.

Since we're discussing hyper-national organisations, I looked at some of the world's biggest charities and causes to see the pervasiveness of greenwashing. I found three significant ways corporate greenwashing has made its way into global charitable organisations. The first is providing 'stamps of approval'. This is where a company, or group of companies, might use a non-profit organisation to rubber stamp some of the bad things they're doing. A great example of this is the World Wildlife Fund (WWF), which produces an annual scorecard[30] measuring how well your favourite supermarket brands deal with palm oil. They measure areas like sustainable palm oil purchasing, supplier accountability, and on-the-ground action. Having one of the world's preeminent animal rights groups rank you highly would be an excellent achievement for any company. That is until you realise all the companies on the ranking are paying members of the WWF's Roundtable on Sustainable Palm Oil. I've worked in palm oil and know the RSPO has done great things. It just seems shady when one of the highest-ranked names is Ferrero, a company that traffics almost exclusively in palm oil.

The second approach is what I'm calling 'the director's chair'. Sometimes a charitable donation is so big, the organisation forgets who's in charge. They end up letting external stakeholder interests dictate their direction. Shell, for example, had issued a gag order when it donated to London's Science Museum.[31] The order forbade the Museum from criticising the sensitive oil giant. There are two other examples that, coincidentally, also feature in my previous book, *Sustainability for the Rest of Us*. The first is about the creation of the UN Millennium Development Goals. While drafted and passed by the United Nations, the Goals were more a tool for powerful blocs and donors to focus direction on what *they* considered the most critical issues—not necessarily what the most critical issues were. Another involves rapper 50 Cent and his work with the World Food Programme in Somalia. Instead of donating food like a normal person, he would only give out a meal for each Facebook like of his new energy drink. It's crazy what a celebrity endorsement will make a non-profit do.

The last approach is the most egregious. It's when an organisation sells their soul by 'signing a deal with the devil'. The Science Museum can no longer be a trusted, impartial source of information given its funding sources. PETA's supposed ties to terrorist organisations call into question all the decades of work they've done to protect animals. Returning to the palm oil plantations of Malaysia, a non-profit called the Orangutan Land Trust receives exorbitant amounts of money from strange bedfellows. Kulim Malaysia Berhad, a palm oil company active in deforestation, gave the Trust US$500,000.[32] They also receive funding from Agropalma, a Brazilian firm destroying the Amazon. I'm not saying the Trust is doing anything wrong with that money. All I will say, though, is that the RSPO dropped human rights abuse charges against Agropalma while the Trust's Executive Director was a member of the adjudicating panel.[33]

While we have been talking about these examples through the lens of greenscamming, it's clear companies aren't even bothering with the scamming part anymore. They've co-opted these institutions that house the world's collective knowledge. They're influencing which global causes are worthy of funding by selecting which can provide the better smokescreen. And they're doing all this while plastering their names over everything. If I were going to get into greenwashing business, this definitely wouldn't be my tactic.

In some ways, though, all these wrong moves are good. Since they're documenting and labelling their greenwashing, it's much easier for you to spot. Compared to some of the *covert* ways companies greenwash at the supermarket, signs for The Adani Green Energy Gallery are pretty in your face.

That ease of identification can lead to an ease of activation. You don't need to patronise some of these places, so don't. If you feel a charity is a front for greenwashing, don't give. Where you do and do not spend your hard-earned money can seriously make a difference. As non-profit organisations surviving pay cheque to pay cheque, your financial decisions might have a more significant impact.

Just ask the Royal Shakespeare Company. In 2019, they abruptly cut short a multi-year sponsorship by BP. They cited dwindling sales, especially among younger segments who were vocal against fossil-fuel funding. In a statement, the company, '... [a]midst the climate emergency, which we recognise, young people are now saying clearly to us that the BP sponsorship is putting a barrier between them and their wish to engage with the RSC. We cannot ignore that message.'[34] Finally, a British institution with some sense.

Chapter 14

Running in Place

I have only the fondest memories of my time at the United Nations. Beyond looking good on a CV and wowing people at cocktail parties, the work there imbued me with a sense of great purpose. It shaped who I am as a person and as a professional. Like any organisation this big and vaunted, though, it has idiosyncrasies and misunderstood parts. Walking through the gates off 42nd Street that very first day, I thought I'd be entering a clandestine world of intrigue, secrets, and the latest high-tech equipment.

Instead, my boss sat me down in a nondescript blue cubicle with a boxy desktop computer running Windows 95.

That, dear reader, sums up the place. It also sums up many of the organisations, governments, and conferences we've discussed throughout this section. Most of the public have a skewed view of these actors, their role in global affairs, and what they can realistically accomplish. For whatever reason, we've come to think of them as far bigger entities than they genuinely are. We view them as an iPhone 14 Plus when they're probably closer to an iPod Nano. They still have a function, sure, albeit a very niche one.

Which is why we've spent so much time exploring how they are no longer fit for purpose. In doing so, we've also raised how they're perpetrating greenwashing in ways not dissimilar to some big corporations. If greenwashing is an act of duplicity, I can't think of a more deceptive segment. On the surface, they go on and on about caring for the Earth. In their back-room dealings, and when the mics are off, they could give a fig about the state of our planet. Unless, of course, they can benefit directly.

We began this exploration by looking at the history of environmentalism at the United Nations; from the creation of the UN Environment Programme, to the UNFCCC, to the Intergovernmental Panel on Climate Change. Going through some of the most consequential global sustainability conferences since Rio in 1992, we found much initial movement but a general stagnation of late. Even worse, today's COP programming is becoming more of a circus in its attempts to attract businesses, influencers, and celebrities along with politicians. Sponsorships, and even chairpersonships, also point directly to greenwashing. All of this indicates either wilful ignorance on the part of organisers and attendees or a format that just isn't working.

To overcome this evolution (devolution?), I gave three recommendations. First, the UN must look to niche down as much as possible. Instead of working to be all things to all people, these events need to be smaller, more focused, and action-oriented. Second, organisers need to find a dragon to slay. Like the successful campaign to address the hole in the ozone layer, having a common enemy makes fighting climate change easier for everyone. Third, there needs to be far more accountability than we see today. Right now, the foxes are running the henhouse.

Leaving the world of the UN, we then found ourselves talking about state-sponsored greenwashing. We discovered how Saudi Arabia's The Line project, while purported to be the carbon-neutral blueprint for cities of the future, is actually a way for a flailing petrostate full of human rights abuses to change its public persona. State-sponsored greenwashing is also happening in places you may least expect it. Australia talks a lot about its beautiful ocean vistas but forgets to mention its penchant for digging polluting stuff out of the ground. Singapore loves to wow you with its gardens but not so much with its insatiable need for oil. Even the European Union hides unsavoury practices. We've learned that zero countries are on track to meet their Paris Agreement targets. So, the mismatch between national image and reality is vast.

Filling in this chasm requires, first and foremost, an understanding by these states that they indeed have a problem. The list of addictions is long. It includes oil and oil lobby money, a weird marriage to old ways of thinking versus far more beneficial modern options, and complacency as politicians only consider the short-term. That's why it's critical that states and politicians prioritise what's important. If the environment or constituents don't matter, be selfish! Divesting from addictions—oil, outdated thinking, complacency—can materially benefit these nations more than they might realise.

Lastly, we addressed the massive amounts of greenwashing happening in hyper-national organisations. These groups don't necessarily operate within the bounds of everyday politics but powerfully shape international policy and direction. The first was international sporting bodies, like FIFA and the International Olympic Committee, and regional players like the NBA. Corporations are using these groups, who gladly accept the cash, to improve their social license to operate. Greenwashing is also commonplace at large meetings of the elite, like Davos, where big talk trumps any action. Perverse intent has also made its way into our most prized cultural institutions. We explored how oil, gas, and other polluting companies are now key sponsors of the world's museums and charities.

Given much of the greenwashing in hyper-national organisations is so blatantly obvious, it's easy for us to spot and call out. I believe we can do three things to push the envelope even further. First, regulate the hell out of who can sponsor what. Like tobacco, there is no reason we should see oil and gas companies having naming rights on stadiums, jerseys, or in any advertising for sport. Next, these institutions should consider alternatives. A permanent home for the Olympics, including more diverse voices at international meetings and more appropriate gallery sponsorship, are all excellent first steps. Finally, if they decide not to listen, make them aware of your dissatisfaction. Sometimes, speaking with the wallet is the loudest voice of all.

While greenwashing is most visible in the private sector, examples from public sector institutions are just as egregious. They convene conferences, ad nauseum, with lots of fanfare but little accountability. They roadshow all their futuristic projects to deflect from projects stuck in the Stone Age. They put their hands out for greedy corporations with dirty money. Then, they dare to take a moral high ground and position themselves as modern-day messiahs.

Most people around the world have come to understand what these actors are playing at. The complicating factor is that there is little we can do to influence these institutions today. Of course, we can vote, choose not to attend a soccer match, or forego Adani's latest 'green' science exhibit. But unless we're in the halls of power writing legislation, or at Davos meeting with the rich and powerful, it's going to be a long time before we see meaningful change.

Which is why it's so, so important to place these institutions in context. It's even more important to remember where meaningful change will come from. As I've said several times, the private sector has the resources and capacity to create a more sustainable future. Plus, they have a willingness none of the organisations in this section seems to have. In the capitalist system we operate within, one silver lining is the fierce competitiveness to differentiate. Without that, corporations cannot survive. Our role is to be the extinction-level asteroid for those companies still acting like dinosaurs. That way, in the fiery words of Al Gore to Davos attendees, '... we can say we are now in charge of our own destiny.'

PART 4

INFLUENCES

Chapter 15

Hey There, Sausage Fingers

Influence.

The word is so commonplace, it's lost a great deal of gravity. We talk about influences as a child, influences in the workplace, and spiritual influences. People ask each other about the most influential things they've ever learned, or the influential quotes they've read. Social media has even turned this once proud adjective into a proper noun: Influencer. These Influencers yield so much power they can rake in big money, assuming they're influential enough.

A whole industry has grown up around influence. Titles like *Getting to Yes, Pre-Suasion: A Revolutionary Way to Influence and Persuade,* and Dale Carnegie's famous *How to Win Friends and Influence People* fill up bookstores the world over. Gurus in the self-help, motivation, and betterment spaces are paid handsomely for speaking to large crowds. Then there are the cute motivational posters of kittens hanging from ropes, jumping in the air, and looking in mirrors with inspiring lines like 'hang in there!'. Go into any human resources department, and you're bound to see them.

But influence is hardly a trivial thing. Our sources of information, from the people we listen to through the media we consume, shape our direction in life. A *positive* influence can give you solid foundations for a successful future. They can encourage you to pursue an education, make a beneficial career move, or get back on the horse after a nasty breakup. *Negative* influences can drag you down, bring out the worst in you, and leave you in a gutter with track marks up your arms. Therapists stake their entire profession on the impact of influence over our lives. Companies pay big bucks to influence what we buy. Religious leaders, jilted lovers, and serial killers take advantage of emotional vacuums to sway our decisions.

There's no better recent example of the impact influence can have than the movie, *Everything Everywhere All At Once*. In the emotional rollercoaster of a film, we follow Michelle Yeoh's character as she navigates the trials of an IRS audit while fending off a collapse of the space-time continuum. You know, just an average Tuesday. Humans have figured out how to jump between different realities in this multiverse. Each reality connects to the others like an intricate spider web. But with each iteration of reality, there are slight variations.

Think of it as similar to evolution. It would be hard to distinguish early humans from their Neanderthal cousins. Put you and a homo habilis man from 2.3 million years ago side by side, though, and I'd bet you could tell who was who. In *Everything Everywhere All At Once*, it's no worries if you're going from our reality to the next one over. Like Darwin's finches on their isolated islands, similarities outweigh differences. But what happens when characters in the film jump to a reality lightyears away? Then the most minor of changes—let's call them influences—grow and expand into massive tsunamis.

In one of the realities, the characters go from launderette owners to movie stars. Another has them existing as rocks. In a third, and probably the funniest of all, humans have evolved to have long, fat fingers resembling hot dogs. This trait renders hands inutile, forcing everyone to use their feet for everyday tasks. There are few scenes in movie history more ridiculous than seeing Jamie Lee Curtis try to woo Michelle Yeoh to bed with floppy sausage fingers, all while playing the piano with her toes.

As outlandish as this is, there are always kernels of truth in art. Philosophers have grappled with the notions of fate, choice, and destiny for millennia. Quantum physics tries to make sense of this through the idea of quantum superposition, the many-universes theory. Does my action of typing this word change the entire trajectory of this book, your day, our lives? Perhaps. On a micro level, I believe influence operates the same way. One influenced choice will compound upon another, so on and so forth, which makes influence exponential. That's why it's so important to keep a vigilant eye on who or what is influencing us. Not only that, but we have to understand how and why they are doing it.

This is a great little general life lesson, of course. But I'm not spending my time here writing a new edition of *Chicken Soup for the Soul*.

In the fight for a better future, some influences have the potential to send us in the wrong direction. Instead of evolving and getting better, we're devolving and doing worse. Like some of the characters in *Everything Everywhere All At Once*, we're watching our fingers go from magical instruments to limp sausages. Unfortunately, that puts every one of us at risk of being greenwashers ourselves. It's likely innocent and inadvertent on our part, but that doesn't make it any less destructive.

Who are these influences, you ask? There are three big segments of sustainability influencers. Not all of them are necessarily negative or hostile. Some are just misguided. Regardless of their approach or intent, each is guilty of exhibiting, promoting, or believing a false narrative. They all say they're green, but a peek behind the curtain shows they're just greenwashing. In this section, we'll explore each of these segments, how they influence the direction of sustainability, and ways we can correct our course into the future.

The first of these segments are **activists** in the vanguard of the sustainability movement. I talk a lot about this group, as I strongly believe they have been terrible representatives of our cause. Maybe fifty years ago, hanging off the side of an oil tanker or joining a protest was good enough. Today, climate activists handcuffing themselves to buildings or blocking city traffic is just a nuisance. They get in the way of progress and do little to enlist more people in the fight for a better future. I'll also walk through some of the various sub-segments, demonstrating why moralists and NGOs can be just as negative an influence.

Next is a much newer but no less influential group of people: the **climate narcissists**. They're an evolution of the social media influencer. For them, it's not about the latest fashions, products, or dance moves. Climate narcissists are using social issues, environmentalism, and the fight against climate change as ways to produce content, gain clout, and make money. We already talked a bit about Kylie Jenner's epic fail with Pepsi, but she's one of many celebrities who make up this segment.

Along with celebrities are the rich and famous, many of whom probably attend those big conferences in exclusive mountain retreats. They throw money at our problems, not realising they are the real issue. Most galling of all are people who don't even know they're in this category. Instead of making a difference, they're just getting in the way.

Lastly, we get to the most critically important influencer of all: **ourselves**. Even though endless influences are coming at us all the time, at the end of the day, we're responsible for how we respond. I get this is a big oversimplification, but I am confident in your ability to make the right decisions. In this section, I'll challenge many of our preconceived notions of building a more sustainable world. We'll walk through some of the biggest misconceptions about saving the planet, look at the lies still persistent in the sustainability dialogue, and examine how a new crop of sophisticated climate denialists risks undoing decades of work. In the end, it's my hope this section, more than the others, will give you the right foundations to become an upright influence yourself.

Chapter 16

Activists, Moralists, and NGOs ... Oh My!

Imposter Syndrome is a common occurrence in the worlds of business and academia. The term is common parlance now that more and more professionals are opening up about their struggles at work. Initially coined by Pauline R. Clance and Suzanne A. Imes in 1978 as 'Imposter Phenomenon', the researchers denoted the syndrome as internal feelings of intellectual phoniness. The idea has since evolved to be a constant, internalised fear of being considered a fraud, even if one is an expert in a particular subject. You undervalue the worth you bring.

Yours truly gets a fair bit of imposter syndrome, especially in new work settings. I like to think of it as an inverse reflection of one's expertise: the more you know about a subject, the less of an expert you think you are. As Miss Badu said, '... the man that knows something knows that he knows nothing at all.'

Every concept, of course, also has its opposite. The opposite of Imposter Syndrome is the Dunning-Kruger Effect. The Effect applies to one who is overconfident in their abilities, especially when they may have none. Most consider it easier to identify versus Imposter Syndrome because those who have Dunning Kruger are typically boastful know-it-alls who can't recognise their shortcomings and have issues with authority. I bet you can think of at least one, but probably many more, people in your life who should have Dunning Kruger tattooed on their foreheads.

Why a psychology lesson when we're supposed to be talking about greenwashing? Our first segment of influences, in many ways, unfortunately falls victim to the Dunning-Kruger Effect. They include the most readily visible representatives of the sustainability movement: activists; moralists; and non-governmental organisations (NGOs). Each has spent decades to rightly attain a leadership position for the cause. They have been central to shaping the sustainability agenda for nearly one hundred years. With this comes power and influence, as well as an oversized impression of what they're genuinely accomplishing.

I'd consider a lot of what they say and do to be the textbook definition of greenwashing. Like the people at Davos, it's a lot of talk but little impact. Unfortunately, this is because much of their approach doesn't fit our current needs. Whether that's giving people the wrong impression of what it means to be a sustainability champion, turning people off by looking down their noses, or petrifying potential supporters, these groups need to rethink their tactics. In this section, we'll explore more of who these groups are, why I think they're greenwashing, and some simple fixes for their outdated strategies.

Activists

Let's start with the group who have been the most influential for the sustainability movement: activists. Over sixty years ago, they were the ones who turned the world's attention towards the environment, animals, and even being better to one another. They were vocal champions at the centre of the cause, staging sit-ins, love-ins, and bed-ins. As such, activists became the poster children for building a more sustainable future.

I want you to close your eyes and bring up the image of a typical activist. They can work in any facet, from environmentalism to animal welfare. It really doesn't matter, because I bet that I can guess precisely what sort of picture you had in your head. Let me know if I get this right. You were probably thinking of some greenie type, like the kind you'd find at a farmer's market or grungy summer festival. They walk around shoeless, the soles of their feet blackened and calloused. Platted, greasy hair likely holds a treasure trove of dead bugs and a stench that could clear the room. Are they hanging off the side of an oil silo or unfurling a protest banner? Maybe you have them flinging fire around indiscriminately from a hula hoop or rope's end. That seems to be one of their favourite pastimes, so it makes sense that you'd envision them this way.

Now, why did most of you think up this same image? There were a million-and-one ways you could have gone. I consider myself an activist, but I mostly wear a button-down shirt and jeans. Bill Gates is an activist, but I don't think he'll be climbing towers anytime soon. Even you, dear reader, are an activist. Did you think of yourself just now?

This is the power of influence at work. At the start of the global environmental movement, these activists represented the majority. Today, though, they are only a tiny fraction of the massive army of people doing good for the planet. Given their historically highly visible position and increasingly vocal tactics, such activists remain what most people see and think of when it comes to sustainability. They endure as the most influential, even if their approaches leave much to be desired.

In my first book, *Sustainability for the Rest of Us*, I recalled a very annoying encounter with your quintessential activists. The story encapsulates exactly who I'm describing when talking about this segment. It also demonstrates the level of influence they continue to have.

It's 6:30 on a wet Friday evening and I'm running terribly late for a movie. Melbourne's traffic can be horrendous during rush hour, but the trams are a pretty safe bet to get you where you need to be on time. Not tonight apparently. We were making great time until we hit the central business district. Here, things just slowed to a crawl. The tram sat between stops for a good twelve minutes. Due to safety regulations, tram conductors can't let people off between official stops. After five minutes the grumbling started. After seven minutes, people started going up to the conductor and asking what in the world was happening. All the conductor knew was there were trams stopped in front of us and nowhere to go. Nothing was coming in off the radio, so she was running blind. By ten minutes, someone got so fed up they pushed open the doors on their own and let a good half of the tram off with them. Within another couple minutes, someone came onto the tram to say all services had been discontinued. We would have to walk the rest of the way. Mind you, it was pouring down rain at this point.

The second I stepped off the tram, I joined a sea of people headed up the main street of the CBD. I mingled amongst the typical tourists taking pictures, and business people on their way home for the weekend. It was certainly more crowded than usual. After about a block, I started to notice a third segment growing more populous by the foot. They grew in tandem with an overt police presence on the streets. Waiting at a traffic light, I turned to the woman next to me. She was soaked to the bone, mascara running down her face, wearing a garbage bag for a poncho. In her hand, she held a makeshift cardboard sign with the Shakespearean scrawling: 'fuck the system, not the planet'. Crap. It was a climate change protest march.

I could see, smell, and hear them now. Angry greenies, vocal university students, and weirdos of every ilk mingling around chanting in unison (kind of). 'Hey hey! Ho ho! The government has got to go! Hey hey! Ho ho! Scott Morrison has got to go!' Those marching down Swanston Street were furious at Australia's federal government for its less-than-stellar response to the country's escalating bushfire crisis. While the stories of bushfire tragedy gripped international headlines, the Prime Minister vacationed in Hawaii. His deputies, beholden to the interests of the coal industry, refused to admit climate change had anything to do with the disaster. Images of blackened forests and walls of fire were made grimmer by the loss of human and animal life. Some estimates point to upwards of 1 billion animals losing their lives, with the emblematic koala added to the endangered species list.[1] The international outcry, and relief, was being met by political silence on the ground.

People had had enough and were demanding heads, and I have to say, I couldn't agree with them more.

But as I pushed my way up the street, narrowly avoiding that many umbrellas jamming into my eye, all I felt was frustration. Not frustration at the ineptitude of the federal government or the tram conductor sitting in her warm, dry seat blocks away. No, my frustration was with the protesters themselves. As they milled about, taking up four lanes of traffic and the entirety of both sidewalks, all I cared about was getting where I needed to be. In the busiest part of the city, at the busiest time of the week, these protesters weren't a force for good. They were just getting in the way.

This frustration only grew as the crowds began to morph from invested, well-intentioned activists to the parasitic hangers-on you often find at events like this. There was the fire dancer trying to get her torches going between downpours; the chanting of the Hari Krishna looking for converts; and, the inevitable Bible basher touting the end times. It was a mish-mash of characters diluting the message at the core of the march. For people on the outside, it was further confirmation that those fighting to save the world were just a bunch of freaks and geeks.

As I finally sat down for the movie, I started to think about all the frustrated people on the tram. Each was just trying to get where they needed to go. Maybe it was somewhere inconsequential, like a movie. Perhaps it was something more important, like a hospital visit or business meeting. Would they look at the protests as something meant to change the world positively? That's highly unlikely. They would remember the protests as a nuisance; something that forced them off the comfort of the tram and onto the cold, damp streets. It made them wet, late, and stressed.

For those with a keen eye, they'd also wonder what all this had accomplished. Weeks later the bushfires still raged, and the government still sat on their hands. There was so much time spent organizing and creating witty banners, tax money used for police patrols and security, and mental capacity thrown into the cause. While certainly coming from a place of passion and good intentions, a simple cost/benefit analysis would have put the protest squarely in the 'unsound' column. Lots of cost, little benefit, and even less impact.

Obviously, these activists had an entirely negative influence on creating change and winning new people to their cause. Worse still, anyone watching from the sidelines will think they must act the same to do something good for the planet. You don't need to compost your waste, forego showering, or twirl fire to make a positive impact. Getting people to think this way does a massive disservice to everyone fighting for change.

The inherent problem with their approach is in how they're marketing the message. Remember, the issue today is no longer getting the message out there. We've spent a century doing that. If people still don't know the world is burning, then I'm not sure what more we can do to show them. Our primary goal today, which we've been quite bad at attaining, is converting people to act. Instead of ostracising people who want to make a difference in whatever way they want, we need to embrace diversity and inclusion for all. That's going to require us finding newer, more modern representations of the movement.

Unfortunately, it seems we're going in the wrong direction. Case in point, the shock-and-awe tactics we've been seeing over the past year. Imagine you're on a family vacation to London. This is something you all have been looking forward to for months. You've been to the Eye, took a boat tour down the Thames, and are now ambling through the National Gallery. There in front of you is one of Van Gogh's most famous paintings, *Sunflowers*. Standing a respectful distance away, you can still make out the brush strokes. Then, *bam*! Out of nowhere, a can of Campbell's tomato soup sploshes across the painting's protective glass. Rather than run away, the perpetrators very brazenly glue their hands to the priceless work of art and exclaim something about feeding the hungry.

What the actual fuck, you think to yourself.

This real-life scenario occurred in October 2022 when members of the activist organisation Just Stop Oil doused *Sunflowers* in soup. This organisation, and others like it, feel destroying art is the best way to alert people to the cause of climate change. They've targeted Vermeer's *The Girl With a Pearl Earring*, Klimt's *Death and Life*, and the most famous painting of all, da Vinci's *Mona Lisa*. When I visited the Louvre in June 2022, security only allowed one person at a time to view *La Joconde*. Even then, you only had seconds before they ushered you out of the gallery.

All over social media, people were rationally asking what any of these attacks had to do with climate change. Their other very public displays of protest, including gluing their faces to the roads of central London, elicited the same curious response. Still, others astutely pointed out the PR blunder doing nothing to galvanise support. I, too, have been asking myself these same questions for years.

Movements like Extinction Rebellion and Just Stop Oil demonstrate how the most vocal are not always the most representative. They also show how something as praiseworthy as stopping the polluting oil industry can be twisted and bastardised until the message is entirely lost. This isn't new territory for activists. PETA is notorious for this type of marketing. Their ads go beyond shock value, showing in all graphic detail just how your clothes are made and your food produced. But after thirty years of this marketing strategy, are any of these ads having the same effect? A study by the University of Queensland measured if using sexualised imagery, as in many PETA ads, sold the same way it does in traditional advertising. In their words, '... is it effective to advertise an ethical cause using unethical means?' They concluded when it comes to ethical advertising, sex doesn't sell. Along the same lines, shock-and-awe approaches do more to turn people off than on

Moralists

Another group who are experts at turning people off are the moralists. Let me set the scene. You're at a mate's barbecue celebrating, let's say, his anniversary. Everyone's having a great time, enjoying the sun, music, and a few drinks. You flit around, ever the social butterfly. Eventually, you mingle with a couple at the far end of the yard. There are the typical vapid conversation points. Where do you live? What do you do? How do you know (insert best mate's name here)?

Tempted by the aroma of smoking meat, you set your mind on fixing yourself a plate of food. You get ready to excuse yourself. Not to be rude, you ask if anybody wants you to bring anything back. Noses scrunched, and eyebrows furled, almost in unison they say, 'Ugh, no, we're vegans.'

Oh, fam. I bet a million little brain cells just exploded in your head. We've all been at a barbecue, party, or dinner where this has happened. And most of you right now are having the same visceral reaction. I bet you're thinking, *that's great for you, but I couldn't care less.* A quick Google search brings up articles like 'Why Do People Hate Vegans' and 'Why Are Vegans So Annoying'. It's clear some vegans have made a name for the entire movement. Like extreme activists, this isn't the influence most probably want.

It's not that I have anything against those who don't eat meat. In fact, reducing our meat consumption is one of the key ways we can help curb climate change. This type of vegan, though, is unique. They are the type that looks down on anyone who would dare think of a steak. They're the loud ones at dinner, proclaiming their admonishment of all meat. Instead of explaining to people why they do what they do, they'd rather boast and make a scene.

Moralists come in all shapes and sizes. It's not just the loud vegan. We have the person in an electric car who gives you a filthy look for driving a truck. There's the judgement at a café when you don't have a keep cup. 'OMG, is that a plastic water bottle?' 'I would never shop at H&M'. 'Yeah, my life is soooooo much better now that I'm living in a van'. It borders on the performative. It reeks of elitism. It doesn't bring people into the fold.

An article in *The Guardian* on why veganism is so off-putting said, '... most people find vegans annoying because it's one of the only social justice causes whose point of entry is entirely negotiated by real, quantifiable, fundamental behaviour change.'[1] I agree entirely. Interestingly, this also supports my notion that many vegans, and other moralists, are greenwashing. If fundamental behaviour change is the goal, which it should be when we're trying to save the planet, they've failed. Not only have they not changed anyone's behaviour, but they've also created an us-versus-them scenario. Even if potentially well-intentioned, their arrogance is not a way to build one's case.

A much more worthwhile approach would be to explain a bit more about veganism, why electric is better, or where to buy a keep cup.

This isn't about cramming it down anyone's throat, but maybe people would be genuinely curious if they weren't worried about being attacked. Rationalising things a bit wouldn't hurt, either. Is it realistic for everyone around the world to go vegan tomorrow, for example? Of course not. As representatives of a cause, in this case animal welfare, getting people to eat even one meat-free meal a week is a win. In the world of sustainability, it's not always all or nothing.

NGOs

> In the arms of the angel
> Fly away from here
> From this dark, cold hotel room
> And the endlessness that you fear
> You are pulled from the wreckage
> Of your silent reverie
> You're in the arms of the angel
> May you find some comfort here.[2]

Are you crying yet?

To this day, anytime I hear Sarah McLachlan's 1997 hit, 'Angel', I think of only one thing: dogs. More specifically, images of abused dogs in shelters. That's because she used this song in a fabulously successful ad campaign for the British Columbia Society for the Prevention of Cruelty to Animals.[3] I re-watched it for the first time in a long time to write this section. Let me tell you. It took a few minutes for me to compose myself.

Audiences in North America are probably familiar with the advert. I think it might even still pop up now and then. For those who haven't had the pleasure, I'll walk you through the two minutes of excruciating sadness. It opens with a black screen reading, 'Every hour an animal is beaten or abused.' Yeah, that's the opening line. Then we go frame by frame with tight shots of bedraggled dogs and cats. 'They suffer alone and terrified.' A matted bichon frisé gives puppy-dog eyes as it looks at you through a rusted cage. 'Waiting for someone to help.' Cut to Sarah McLachlan and a cute golden retriever imploring viewers to save a suffering animal from its abuser. Cue happy, playful animals. Then one parting shot of a rescue dog's big, brown eyes piercing deep into my soul.

If that doesn't move you, it's time to check your heart. Nearly two decades on, many people worldwide can still instantaneously associate the song, video, and message. That's pretty good staying power if you ask me. The unfortunate part, though, is people's reaction to the ad. Sure, the BCSPCA raised a lot of money from the campaign. As I hold back tears, my first thought isn't to give to the charity. Instead, I reach for the remote to stop the video. Honestly, I don't think I've ever been able to make it to her call to action. Even Sarah herself has told the media she can't watch the entire ad. For most audiences watching, still today, they'll never make it through either.

So is the ad having the intended effect? Undoubtedly, it drives an emotional response and leaves a lasting impression. It also raised over US$30 million for the American Society,[4] their biggest campaign ever. While notable, after two years the money stopped coming in. Perhaps the Society moved onto other things, or people found other causes to care about. My sneaking suspicion, though, is that audiences became tired of the sad-vertising.

Sad-vertising is exactly what it sounds like: sad advertising. Sarah McLachlan's ad might be among the most well known, but it isn't alone. Typically put together by non-governmental organisations championing social causes, they pull on your heartstrings to open up your purse strings. Sustainability professionals, especially those in the social-good space, are the biggest pushers of sad-vertising. We've been using it for decades. Think about the World Vision ads of the 1980s, where all it took was ten cents a day to change a child's life in the developing world. It's not like the cause itself wasn't sad enough.

We needed to take it and overlay a sombre mood, light folk music, and sepia tones.

Mothers Against Drunk Driving has an ad where a car flips multiple times over a fence, into a backyard, and onto a child. The father holds the mangled body while the drunk driver looks on helplessly.

A Singaporean Government anti-smoking PSA shows some pretty gruesome videos of people having their legs amputated, trying to talk without jaws, and losing fights against cancer.

And don't forget that sad, sad polar bear on the melting ice cap. I wonder what ever happened to her?

Sustainability has generally been pretty bad with how we've done advertising. I would have loved to have been in the room when whatever genius thought up modern-day sustainability messaging. The burned-out hellscape of an Earth brought to its knees by humanity's misuse of natural resources. Desert vistas covered in the red dust of a nuclear winter. Humans, somehow clad in black leather jackets, race around on motorbikes to find what they can to survive.

If we want to get our message across, we have to do better. Sadvertising just isn't it.

For the corporate world, using emotional advertising has been shown to increase sales. People tend to buy things emotionally, so this checks out. In the non-profit and social-good world, however, studies have shown the opposite to be true. Brave, an ad agency based in the UK, wanted to see whether emotional advertising worked with your stereotypical charity ad. They used biometric technology to gauge and measure reactions. An advert from Take Water Aid scored the lowest.

> Using face-tracking, the most prominent emotions our cohort expressed were disgust and fear. Measuring the strength of emotional response through galvanic skin response, we saw spikes in emotion when footage of a newly-installed well with smiling children drinking clean water was shown. But when the advert flicks back to children drinking dirty water, that positive response disappears.[5]

Adverts with a happier message or more positive imagery scored higher, close to double those using sad-vertising. Add to this the sheer saturation of sad-vertising coming from NGOs. When there's too much of something in public view, audiences tend to ignore it. Depress or disgust them enough, and they'll turn it off completely. People in the space diplomatically call it 'compassion fatigue'.

As is their m.o., the guys at *Southpark* took all this to a most ridiculous level in an episode titled, 'Safe Space'.[6] While shopping at Southpark's brand new Whole Foods Market, Randy eventually makes his way to the checkout register. Reaching into his wallet to pay, the cashier asks if he wants to donate a dollar to help hungry kids. He meekly responds with a no, to which the cashier does a doubletake. Then Randy's made to confirm all this on a touchscreen. When the machine doesn't work, the cashier says it's usually never a problem because everyone always gives a dollar.

Exasperated, Randy says he does give to charity. He just doesn't want to every time he comes into the store. Then the cashier guilts him further by making him speak into the store microphone so everyone can hear. Embarrassed, he shuffles out. The next day, it happens again, except this time Randy needs to touch an image of a child's distended stomach to confirm. To get his change, he's made to pull a sandwich from a cardboard cut-out of a hungry child. When it gets stuck, the cashier says, 'Oh yeah … she's a hungry one.' With his foot on the child's face, Randy is finally successful. Randy leaves angry at the cashier, his day ruined by the humiliation.

While our local store may not make us go to such lengths, the sad-vertising and guilt they pile on us are no different.

Marketing 101

I want to be clear in my unwavering support for these influential groups. The modern environmental movement might have never gotten off the ground if it wasn't for them. It's because I care that I want these leaders to do better. Activists lack pragmatism. Moralists lack empathy. NGOs have sold out. This has leaked into messaging, voiding the influential leadership positions these groups claim to have.

If they continue to be on the frontlines, loudly squawking on the street, they need to get the message out there better. This isn't just about scale but also impact. They need to become more pragmatic, patient, and purposeful to regain trust and true influence. My recommendation is that they begin to think less like activists and more like marketers. They must return to marketing basics, the most fundamental of which is remembering the audience. One must know what energises them and what puts them to sleep; what gets them talking and what makes them cringe. Without these critical pieces of intelligence, you risk losing your audience along whatever journey you're trying to take them on.

Right now, these groups are all putting their audience off. Rather than developing messaging or strategies around what their audience wants, many are trying to force their wishes on an unwilling market. Whether it's vandalism, snootiness, or sad-vertising, people are tuned out. In short, they've forgotten what their audience wants, and it's starting to show.

Instead of explaining why this is all so important, it's much easier to show what happens when you forget about stakeholders. Take, for instance, the example of Tony Piloseno. A young college student, Tony was passionate about paint. He channelled this passion into a job with one of America's oldest and most well known paint brands, Sherwin-Williams. Tony took his love of paint to TikTok and amassed millions of viewers to his paint-mixing challenge videos. These videos were a viral sensation for people stuck inside their homes during the pandemic. Sherwin-Williams' management didn't see things the same way. In fact, they fired Tony for breaching the company's (outdated) social media policy.

Had Sherwin-Williams approached this more strategically and

truly understood its stakeholders, the company could have capitalised on the apparent desire for information from a non-traditional channel. Instead, they ended up a social media laughingstock and potentially lost an entire generation of consumers. Don't worry, though. Tony landed on his feet and secured a job doing what he loves at Sherwin-Williams' biggest competitor.

For activists, moralists, and NGOs, the benefits and risks of knowing an audience are no different.

The next logical question would be, *okay … but how do I identify my audience*? You need to truly understand what I call the *universe of stakeholders*. These are all the possible segments that might resonate with your message and who you want to court. Luckily when it comes to saving the world, we have a few significant segments to choose from. The first are those who already support your cause. That doesn't mean they have to be holding a megaphone or carrying a banner. It means they understand, believe, and do what they can. When dealing with this segment, messaging should inform them of the latest news and give them channels to keep contributing as they see fit. On the opposite end are the lost causes. These are people who will never support what you're doing. In fairness, maybe they will help one day. But it will take a lot of effort to get them there. With limited capacity, I'd recommend putting them to the side for now.

Then you have the third segment: people who want to support but don't know how. They are the biggest of the three, so plenty of opportunities exist. The key is to keep it simple. They need easy-to-understand information about why your cause matters. For extra credit, include why it should matter to them. It would help if you also told them *how* they could help, which can be through giving their time, expertise, or resources (AKA money).

At this stage, don't overthink things. Remember, we're getting back to basics. If all you remember is to be more pragmatic, patient, and purposeful, you're already in a much better spot than where you started.

For those of you who aren't activists or in an NGO, I apologise for the sidestep into the world of marketing. I genuinely believe these groups must change their approaches and message, especially with this current incarnation they've taken. Continuing in this way will not only block new support but also begin to haemorrhage the supporters we already have. It will also make our already difficult job even more challenging as we have to explain away the damage they've done.

Like I said at the very beginning of this chapter, activists, moralists, and NGOs may be the most vocal groups out there. They are not, however, representative of the wide world of sustainability. If you're reading this and think saving the Earth means you have to live zero-waste, shop at a second-hand store, or never get on a plane again, you're wrong. Not to wax poetic, but if we're going to face the challenges ahead of us, then we need all of humanity fighting alongside. That means you, no matter who you are, where you come from, or what you care most about. There's plenty to do and plenty of room to welcome everyone.

Chapter 17

Bright Lights Cast Long Shadows

Celebrities and social causes. It seems like the two have always gone hand-in-hand. Coco Chanel sought to empower women through her designs. Jane Fonda campaigned against the Vietnam War. Princess Diana worked tirelessly to destigmatise HIV and AIDS.

For a brief period in the mid-1980s, big gangs of celebrities would gather together for this cause and that cause. The most famous of these is probably the 'We Are the World' campaign for the charity USA for Africa. Nearly fifty of the decade's biggest pop stars sang their hearts out on the catchy song, which still ranks as one of the top singles of all time. A group of British musicians by the name of Band Aid, among them George Michael, Sting, Bono, and Boy George, released 'Do They Know It's Christmas?' to raise money for the Ethiopian famine. Radios, malls, and grandmas all around the world still play this jingle every holiday season.

Not every campaign's going to be a chart-topper. In an earlier chapter, I talked about 50 Cent's work exploiting starving kids for likes. It doesn't end there. Pre-pandemic, the Norwegian Students' and

Academics' Assistance Fund would give out an annual prize called the Rusty Radiator Award. This would go to charity campaigns that didn't quite hit the mark. Some of the biggest flops were those with celebrity ambassadors. The 2017 grand prize went to British charity Comic Relief with their campaign, 'Ed Sheeran Meets a Little Boy Who Lives on the Streets'.[1] I wish that was a joke.

Over the course of a five-minute video, we see Ed wandering the streets of Liberia's coastal capital, Monrovia. We faintly hear one of his newest songs in the background while he talks about being 'on edge'. Eventually, he meets a little boy named JD, who is homeless and orphaned. They chat. Ed packs up to leave. Then, he feels so guilt-ridden he needs to bring a very big group of people together to announce he'll pay to put JD in a hotel, no matter the cost. Ed explains to the viewer that the point of this video was to pull on their heartstrings. But he's so magnanimous he couldn't help himself.

Intentionally or not, it turned into a video about how good a human Ed Sheeran is. The jury said the video was poverty porn, devoid of dignity and outdated. It never addressed the core issues at play in Liberia, like poverty and corruption, because it spent most of the time addressing Ed Sheeran's feelings.

The runner-up that year was an appeal by Tom Hardy to feed starving kids in Yemen. His approach: lean into the camera, wipe a tear, then show graphic images of children near death and guilt the public into donating. A resurrection of Band Aid, this time with new pop stars and fundraising for Ebola, snagged the 2015 award. Judges said the campaign '... contributed to the spread of misinformation and stereotypes of Africa as a country filled with misery and diseases ... furthermore, they just make it so much more about themselves!'[2] Seems a pretty good summation of the problem we're about to dissect. Overall, though, these latter examples were mostly inconsiderate and relatively harmless. While embarrassing for those involved, and uncomfortable for us to sit through, a charity got some money and people went on about their lives.

Compare that to today's celebrity do-gooder, who is a different breed entirely. I want to preface all this by saying I'm painting with broad strokes. There are, of course, lots of celebrities using their platforms for good. Notable activist celebrities include George Clooney, who does wide-ranging humanitarian work in Africa and the Middle East. Angelina Jolie represents the UN as a Goodwill Ambassador, works directly with the UN refugee agency, and has even bought land in Cambodia to preserve the country's wildlife. Shakira's foundation builds schools in Colombia. Jim Carrey advocates for men's mental health. Oprah Winfrey uses her considerable wealth to better childhood education.

This isn't about them.

It's about the celebs—and I'm using that noun very loosely—who use social and environmental issues to deflect away from some fairly unsustainable behaviour closer to home. Like corporations, these celebrities apply a liberal coat of green sheen to their public image. If I'm feeling generous, I'd say they do it out of ignorance. Let's call it accidental greenwashing. Though, with their teams of agents, assistants, and reps, I doubt very much that all of it is blameless.

Beyond this, I believe many of them think they're actually doing something good. They've convinced themselves the positive work they do in one realm offsets all the not-so-positive stuff in other parts. Just like a thirty-minute workout once a week doesn't entitle you to a thick shake, the benefit these celebrities provide doesn't come close to absolving them of the costs they're inflicting.

This next part of our journey together will take us from the Mediterranean coast, to the Egyptian desert, to the stratosphere. We'll go from the heights of wealth to the depths of egoism. Along the way, it will become clearer and clearer how the ultra-wealthy, ultra-influential, and ultra-narcissistic are remaking greenwashing in their own image.

The Ultra-Wealthy

One of my favourite places in all of New York City has to be the Rose Reading Room at the Public Library. It exudes the elegance of a bygone era. Stretching nearly two city blocks, the room's eighteen tiered chandeliers glimmer off the rich parquet floor and mahogany desks. Giant arched windows line either side, illuminating the stacks of leather-bound books. Almost five storeys above you is the room's defining feature—its ornate ceiling. The New Yorker describes it best in saying, '... [e]ven by the building's Beaux-Arts standards, there are a lot of gilded curlicues and cornucopias and flute-playing cherubs cavorting up there, around celestial murals of soft pink clouds.'[3] When you're there, you know it's someplace special.

Even then, it's just one of the many jewels in this fabulous building. The main branch of the New York Public Library, the largest public library in the largest city in the United States, houses many wonders. Named after the founder of Reader's Digest, the DeWitt Wallace Periodical Room murals detailing the city's history line the walls. The Library's map division is one of the most extensive in the world. Before you exit the building, you walk through Astor Hall. Its marbled staircase and vaulted porticos are reminiscent of a cloister more than a library. Stepping outside, the grit and noise of the streets bringing you back to the modern day, the City's most famous statues stand guard as they have for over 100 years. Patience and Fortitude, the Library's two emblematic lions, bid you farewell while tempting you to return soon.

Interspersed among New York's modern skyscrapers, neon billboards, and cluttered bodegas are testaments to man's desire for permanence. Massive structures, built a century ago for the public good, dot every street and avenue. They feature in classic films and on tourist junkets. Like the Library, they're just as grand and imposing as anything you have to crane your neck to see. From the Bronx to the Battery, Wall Street to 42nd Street, these testaments are an integral part of the city.

Some you may know, like the Metropolitan Museum of Art, the Natural History Museum, and Carnegie Hall. Others, a little more obscure but just as noteworthy, include the Frick Collection and Morgan Library. These nineteenth-century public works were financed by some of the wealthiest people who ever lived. Their family names are a who's who of the well-to-do: Rockefeller, Astor, Vanderbilt, Carnegie, and Morgan. Some called the patriarchs of these families robber barons, given their ruthlessness in business. Others, like the Astor family, had intergenerational wealth stretching back to the foundations of America. In the elitist atmosphere of the Gilded Age, where there was no such thing as too much decadence, never has so much been held by so few.

To put things in perspective consider some of the *crème de la crème*. Cornelius Vanderbilt, one of the wealthiest people at the time, had more money than the US Treasury. This is nowhere near the most monied person in modern history, Andrew Carnegie, who had an eye-watering US$446 billion floating around.[4] That's more than twice as much as Elon Musk and nearly four times more than Jeff Bezos, both of whom we'll come to shortly.

While the robber barons of the Gilded Age unquestionably did become spectacularly wealthy, they had to give back to be considered part of proper society. With so much wealth, one could do practically anything. Environmental concerns were not front of mind at the time. So these billionaires spent their money on the public good. Each gravitated towards culture and the arts, education, and healthcare. They set up foundations, many of which still do excellent work today. We've listed some of the grand buildings placing New York City on the map as a cosmopolitan cultural centre. Other projects included water works, public housing, and poverty alleviation. The Rockefeller Foundation notes the rich of the day '… were among the small group of individuals who created modern philanthropy by attempting to deal with the root causes of poverty, disease and ignorance rather than simply ameliorating their symptoms through charity.'[5]

Was all this work entirely selfless? Not by any stretch of the imagination. I mean, people called them robber barons for a reason. These men were looking for ways to slap their name on something altruistic while probably getting kickbacks in return. In that respect, they were very successful.

It's no different for the ultra-wealthy today. They're all still looking for ways to leave a mark on society. But the 1800s are long gone. There are rules and regulations when it comes to business. Today's issues are global, not localised in Manhattan. Plus, they are far more existential than housing the poor or educating the young. Today, we're talking about the collapse of society and the destruction of the environment.

I'm of two minds when it comes to the modern, ultra-wealthy philanthropist. On the one hand, you can't deny the positive impact vast sums of wealth have when used appropriately. *Barron's* estimates the world's wealthiest gave close to US$200 billion to philanthropic causes in 2020.[6] That's a lot of money that can go to research and programming. On the other hand, there's that nagging issue in the back of my mind. It's the same issue that's come up again and again in researching and writing this section of the book. What are their true motivations? Why are they doing what they're doing? Does it all offset their less-than-admirable qualities and actions?

Of course, I'm no psychologist. Even if I was, it's doubtful I'd get Elon lying on my couch to examine his psyche. What I can do is compare and contrast deeds versus actions. Thus, I found there are three unique clusters of ultra-wealthy philanthropists. These match up quite well with the groups of corporate actors we've discussed. The first are those that do pretty well walking their talk. Next are those whose words don't always match their actions. Their mistakes are often made innocently or out of ignorance, not malfeasance. Then there's a group who have a lot of explaining to do. With them, there is an egregious mismatch between claims of trying to help humanity with outright actions to destroy it.

Off the coast of Seattle, in the foggy waters of Puget Sound, lies a private island only accessible by seaplane. According to those who've researched the location, the owner is a bit of a recluse. He has a luxury mansion tucked away amongst the island's forest of red cedars, and a series of bunkers deep underground. It's in these bunkers he conducts clandestine work far away from the public's prying eyes. Since acquiring his billions, sources say, the man has been on a mission to implant microchips in every person on Earth. Some call it the Mark of the Beast from Revelation, while others think it's a way to control our thoughts and actions. This man was also responsible for the Covid-19 pandemic, working with the Chinese Government right here on the island to manufacture the virus. Bill Gates, the founder of Microsoft, is so wealthy there's no stopping his evil grasps for power.

Take the tin foil off your heads. Although it might seem ridiculous to any sane person, these are genuine conspiracy theories that millions have about multi-billionaire Bill Gates. Who knows how or why they started or why they're so pervasive. Maybe they are all true, and I'm the crazy one not paying attention. Even so, perhaps a little microchip in the arm is worth it given all the money he and his Foundation spend on social causes.

Apparently, at the same time as trying to take over the world, Bill's also able to give out billions in charity through the Gates Foundation (where does he find the time?). With an endowment of over US$50 billion, the Foundation works to address some of the world's most critical social issues.[7] Since the start of the pandemic, they have committed over US$2 billion to the Covid-19 response. In 2020 and 2021, the Foundation gave over US$6.5 billion each year to nearly 2000 global projects. All told, since the Foundation's creation in 1994, they have made US$65.6 billion in grant payments to worthwhile causes. It is among the largest, if not the largest, foundation in the world.

If we get philosophical for a second, what makes someone charitable? Is it how much they give in total, or how much as a part of their income?[8] You could compare the percentage of giving to total wealth. Warren Buffett, founder of mega-conglomerate Berkshire Hathaway, is a notorious giver. He gives out billions in company stocks each year, including an ongoing endowment to the Gates Foundation. Along with the founder of Duty Free Shops (DFS), Chuck Feeney, both men have made very public pledges to give 99 percent of their wealth to charity. Feeney, known as the James Bond of Philanthropy, had to shutter his Foundation in 2020 when it had given away all his money. He now lives in a rented apartment in San Francisco with his wife and ten-dollar Casio watch.

One can also calculate charity based on the total dollar amount of giving. George Soros, another philanthropist suspected of bringing about the end times, uses his Open Society Foundations to give billions towards countering authoritarianism and encouraging an open press. My fraternity brother and broadcast mogul, Ted Turner, gave so much to charity he was taken off the Forbes Rich List. His giving includes US$1 billion to establish the UN Foundation, a group aimed at raising philanthropic money from big investors.

Some spend their dollars closer to home. Michael and Susan Dell, of Dell Computers, support poverty alleviation across the United States and in Austin, Texas, in particular. Former Microsoft CEO, Steve Ballmer, works to combat mental health issues in Washington state. Pierre Omidyar, founder of eBay and PayPal, invests billions through the Open Markets Institute and Public Knowledge. These non-profits advocate for stronger antitrust laws and are critical of big tech's role in society. It doesn't get less self-serving than taking down the very industry you work within.

No matter which way you slice or dice it, the world is fortunate to have a multitude of ultra-wealthy philanthropists using their wealth commensurate with their status. Beyond the name recognition, it's hard to connect any dots where the giving from these individuals is overtly untoward. Unless, of course, you count Gates giving to Covid-19 research as a way to get his chip in every arm around the world.

Then there are the ultra-wealthy who give but trip up along the way. It's not like they're being two-faced or deceitful. I believe they do have the best intentions in mind. But through ignorance, oversight, or having an awful PR manager, these rich make themselves easy prey for scrutiny.

Like when they travel to important climate events. In 2019, international media lambasted celebrities attending Google's green summit in Sicily for arriving in an armada of 114 private jets and a flotilla of superyachts. These aren't exactly the most sustainable modes of transportation. British tabloid, *The Scottish Sun*, labelled Prince Harry 'his royal hypocrite' for showing up to the Google event via jet. Climate activist and director Leonardo DiCaprio came on his mega-yacht. Even Katy Perry, the UN's Goodwill Climate Ambassador, saw nothing wrong with flying private to Sicily. These silly micro-aggressions cause the public to sit back and wonder what the hell these people are thinking.

They're not thinking. That's the answer. It's not that their actions are entirely innocent because there are indeed dire repercussions. Although you can say, 'Whoops, sorry', when the media attacks you and your jet, the damage has already been done. Yet I'm inclined to forgive these lapses in judgement and chalk them up to expensive learning experiences. Such actions don't necessarily label these individuals greenwashers, but we should rake them across the coals if they make the same mistake twice.

What floors me is how a charitable individual like Bill Gates gets pilloried as the Antichrist. Media ridicule a vocal spokesperson for environmental conservation, like Prince Harry, as if he were a child. Then, people idolise someone who very much resembles and acts like an actual Bond villain. I'm talking, of course, about Amazon founder Jeff Bezos.

The issue with Bezos, and equally with Elon Musk, is the cognitive dissonance their actions invoke. This was on full display with the launch of the Bezos Earth Fund in 2020. A US$10 billion fund for (buzzword alert!) transformational systemic change, the whole thing reeks of greenwashing. First, let's address the elephant in the room: how Bezos made a fortune destroying the planet. Amazon's Scope 1 carbon emissions are staggering, growing by 18 percent between 2020 and 2021.[9] This doesn't consider emissions from all the shops supplying Amazon's products, which the company doesn't bother monitoring. His infamous trip inside a giant dick to promote Blue Origin and space tourism also had negative impacts, which are impossible to calculate fully. When pressed, Bezos doubled down on the greenwashing. He stated the mission of Blue Origin was to move polluting industries off the planet. How benevolent!

There are also inherent issues in the set-up of the Fund and how it distributes its grants. The biggest of these is in grants going to established non-profits, primarily in the global north. *Curious Earth* notes only two of the grantees are non-white-led, calling into question whether those most impacted by climate change will really benefit. Having established organisations receiving massive amounts of funding also seems to perpetuate the vicious cycle we're already in. Why would we give them more money if these groups have been around for half a century and still don't have the answers? Wouldn't it be better to invest in newer, more innovative, nimble non-profits that may have the answers we've been missing? As it stands, none of this points to anything near transformational change. California-based non-profit, Center for Community Action and Environmental Justice, put it plainly:

> If the Earth Fund wants to purport to save the planet, they should send funds directly to grassroots communities who are the least responsible and hardest hit by climate disaster and the kinds of rapacious business practices Bezos engages in.[10]

Then we have every incel's favourite douche, Elon Musk. He's rich, but it's not like he gives lots of money to environmental organisations or social charities. What he *does* have is a reputational stranglehold on being a climate innovator, primarily because of his work in electric vehicles. Tesla has undoubtedly upped public adoption of EVs, but that doesn't mean Elon's necessarily gung-ho on the idea of saving the Earth. Musk was livid when the S&P 500 dropped Tesla from its ESG Index, noting unethical practices and a general lack of transparency. He described ESG as a 'scam' that's been 'weaponized by phony social justice warriors'.[11] Maybe that is why he doesn't let Tesla track its emissions or disclose how it handles waste. Nor does it talk about being ranked one of the world's top hundred polluting companies, given its years of violating the US Clean Air Act. If ESG's a scam, I guess it makes sense to ignore evidence of unfair labour practices and systemic racism in the company, too. Gosh, Elon's just so on it.

Bezos and Musk aren't the first rich folks to throw money at an image problem. Those wealthy New York men building monuments to their power wrapped this egoism in a blanket of public good. Today, the same is happening with the hottest public good possible: saving the planet and its people. The Chan Zuckerberg Initiative supports criminal justice reform and bio-medical research. Never mind Facebook actively foments violence, like with the genocide in Myanmar, and uses unethical means of manipulating human emotion. Facebook also continues to profit from climate denial funded by the fossil fuel industry.

The Schusterman Family Foundation gives billions towards Israeli causes and gender equity. They rarely mention all that money comes from the late oil billionaire's Samson Investment Company. Oh, and they also forget to tell people a lot of their philanthropic dollars go to pro-settlement, far-right, and anti-LGBTQIA+ groups around the world.[12] What a strange interpretation of gender equity.

Charles Koch, arguably one of the vilest creatures to ever roam the Earth, founded the charity Stand Together. Informally known as the Koch Network, this group focuses on poverty alleviation, criminal justice, and education. Interestingly, in his pursuit of creating an

American plutocracy, these are the exact same areas he funded Republicans to rail against. His philanthropic investments also go to organisations opposing environmental regulation. Those groups include the Heartland Institute, Pacific Legal Foundation, and the American Institute for Economic Research. His money also went to groups influential in President Trump's decision to withdraw from the Paris Climate Agreement.[13] *The Guardian* named Koch one of America's top climate villains.

An honourable mention also goes out to philanthropist couple Karen and Jon Huntsman. Jon initially saw business success developing the world's first plastic plates, cutlery, and fast food containers. He then reinvested all his money into the Huntsman Chemical Company. The Company specialises in the manufacture of consumer and industrial chemical products. This is where he made his fortune. He would survive a prostate cancer diagnosis in the early 1980s, prompting him to found the Huntsman Cancer Institute. Over the next forty years, the Institute would grow to become one of America's major cancer research centres. Huntsman would get cancer three more times before his death.[14] The irony, which shockingly went over the family's heads, is in the link between cancer and the chemical industry he championed.

Professor of organisation studies at the University of Technology Sydney, Carl Rhodes, takes issue with mega-donors with incriminating backgrounds: 'The people who attempt to make these changes through philanthropy would never admit that they're part of the problem. Instead, they present themselves as messiahs coming to solve it. It's hubris in the extreme.'[15]

I guess the robber barons never died out. They just got greedier.

In no way am I advocating for the ultra-wealthy to give away all their money to save the planet. If they want to, that's great. They've worked hard for what they have and should spend it how they see fit. Unfortunately, giving a large proportion to social and environmental causes isn't realistic, equitable, or something we should bank our hopes on. We can't expect the efforts of a few individuals, no matter how admirable or vast, to dramatically change things for ten billion people. Even those ultra-rich who have given away all their money, like Feeney, only nudge the needle in the right direction. Creating the transformational systemic change Bezos claims he wants will take more than a few billion dollars.

Celebs

Hollywood sits smack dab in the middle of one of the world's hippie hotbeds, Southern California. Prius-driving vegans wear ethically sourced hemp clothing. A dozen types of plant-based milk are available at your local store. You can even get an OCal certification from the government to show the weed you're farming is organic.

Rich and famous folks sitting up in the Hills don't want to be left out. They're just as adamant about having a green lifestyle and saving the planet. That's why they go to luxury markets like Erewhon to buy Hailey Bieber's US$17 strawberry smoothie, a 1.5-litre bottle of hyper-oxygenated wellness water for US$25.99, or a 25-gram pack of vanilla powder that retails at US$65. That last one says it's '... grown with integrity, gives back to Pacific communities.'[16] Obvi celebs are so doing their part, too!

During their annual celebrations of each other—AKA awards season—the A-list crank up the activism. The world's attention is on La-La Land, so why not capitalise on the opportunity? Historically, shows like The Oscars and Grammys have been where actors call for change. They're where #metoo gained wide public attention, and calls for better diversity in the arts began. In 2020, the universe must have manifested something big because the *cause du jour* was environmentalism.

All the big shows played ball. Instead of having brand-new couture dresses or bespoke suits made for the event, organisers of the Academy Awards encouraged attendees to adopt more sustainable fashion. Actors re-wore previous red carpet attire, upcycled from other designer dresses or purchased frugally from eco-conscious brands. The Golden Globes were an entirely vegan affair, with glass replacing plastic at the venue. Even Britain's version of the Oscars, the BAFTAs, joined in. Along with an all-vegan menu, there were no swag bags, and the red carpet was made of recycled material. Guests even received a helpful guide from the London College of Fashion on how to dress sustainably.

I'd love to think celebs are doing things like this as a way of using their platform to spread a positive message. With millions of eyeballs watching these awards shows, there's no doubt the message will get out. My inner cynic tells me the real drivers are either chasing clout and staying relevant or assuaging guilt and absolving transgressions. Perhaps it's both? Their every action is under a public microscope. That means any dissonance between what they say and what they do will undermine the merit of their campaigns. It only took a little tugging to see the whole thing unravel.

The red thread was the juxtaposition of these infinitesimally small, out-of-touch efforts against the massive juggernaut of climate change. Media and peers heaped praise on celebrities like Joaquin Phoenix, who re-wore the same Stella McCartney tuxedo to each award show. 'What bravery,' McCartney said. Laura Dern donned an Armani dress she last wore in 1990. Saoirse Ronan engaged the House of Gucci to use leftover fabric for her red-carpet look. Penelope Cruz had on couture Chanel from 1995.[17] Mind you, a lot of these pieces have only ever been worn once. Personal assistants had to dig them out from the back of a warehouse somewhere in Burbank. I mean, what *won't* these people do to save the Earth?

Well, they won't give up their swag bags. Even though they may have slummed it with Michelin-starred plant-based foods, those at the Golden Globes also gobbled up gift bags worth nearly US$10,000. Along with spa packages, personal vibrators, and kombucha, guests received a seven-day trip to Hawaii.[18] Wait, what? Flying economy from LA to Honolulu emits about 660 kilos of carbon. That's what an average person uses over four months.[19] Going vegan, the part making headlines at these events, saves less than 5 kilos per meal.[20] Doing the math puts the celebs at a 655-kilo deficit if they take that trip.

These rich and famous don't fly coach like the rest of us, though. They love flying in their private jets (which Joaquin has begged them not to use, by the way). Long story short, private planes are exponentially worse for the environment than your typical A380. Instead of shuttling upwards of 300 people at a time, private jets are, by nature, private. Studies have found '... a single private jet can emit 2 metric tons of CO_2 in just an hour ... the average person in the EU produces about 8.2 tons of emissions over the course of an entire year.'[21] They're less fuel efficient, especially compared to more eco-friendly transport like cars or bikes. Colin Murphy, deputy director of the Policy Institute for Energy, Environment, and the Economy at the University of California at Davis noted, '... [a] fully loaded 737 has about the same emissions per passenger mile as an efficient car like a Prius.'[22]

A firestorm erupted on social media in 2022 after Kylie Jenner posted the now-deleted Insta pic of her jet, her partner's jet, and the caption, 'You wanna take mine or yours?' TikTok blew up, and Twitter users started tracking celebrity flying habits. A UK-based sustainability firm then analysed all these little nuggets of info to identify the biggest climate criminals amongst the glitterati.[23] Swifties weren't happy their girl Taylor came out on top, with her jet making a whopping 170 flights in the first half of the year. The emissions from those trips equalled 1184.8 times more than a normal person's total annual emissions. Other names on the list included Jay-Z, Spielberg, Kimmy K, and Marky Mark. Particular anger went to Drake, who took a fourteen-minute flight, and Kim with her twenty-three-minute San Diego jaunt.

On average, the study revealed these celebrities emitted 3,376.64 tons of private-jet CO_2 during the first part of 2022: 'That's

482.37 times more than the average person's annual emissions.'[24] It's not like these were long-haul, either. Typical flight times were just over an hour, covering around 100 km per trip. Yes, you read that right. It took these celebs just as long to fly as it would have taken them to drive.

Drake showed just how out of touch he was with the public by responding, '... [t]his is just them moving planes to whatever airport they are being stored at for anyone who was interested in the logistics ... nobody takes that flight.'[25] Great. So all that damage is being done with nobody even on these planes.

Not only that, but filmmaking is an extremely polluting industry. It takes thousands of people to create your favourite blockbuster. The set design, travel, fuel, and everything else involved add up. A 2021 report found the standard carbon footprint for one of your mega-films was 3370 metric tons.[26] This is up dramatically from a similar 2006 study, which calculated a 500-ton footprint.[27] None of this accounts for the long list of other sustainability-related areas that need fixing in Hollywood, like closing the gender pay gap, equal representation, and responsible consumption.

In fact, the 2021 study isn't even fully representative. It includes self-reported data from ten studios out of dozens working in California. Further, numbers are only what sustainability professionals call 'Scope 1'. These are the figures directly related to what a business controls. They don't account for Scope 2 or 3, which are things like the actions and performance of a supplier. Product placements don't count, either. So when you see Sony come up in *Spiderman* or Audi in *Avengers*, the studio takes no responsibility for rare-Earth mineral mining or fossil fuel consumption.

I understand the job of actors, producers, and directors is to create worlds of fantasy. This is all taking it a bit far, no? Maybe 2020 Golden Globes host, Ricky Gervais, had it right in his opening monologue.

So if you do win an award tonight, don't use it as a platform to make a political speech. You're in no position to lecture

the public about anything. You know nothing about the
real world. Most of you spent less time in school than
Greta Thunberg.

So if you win, come up, accept your little award, thank your
agent, and your God and fuck off, OK? It's already three
hours long.[28]

Then there's another issue entirely. It seems some celebrities have a very
different interpretation of green than that of environmentalists. For
this group, the potential goldmine coming from brand endorsements
is too tempting to pass up. On the flip side, companies are salivating at
the chance to get a big-name celeb to represent their product. They'll
throw whatever money it takes to sign 'it' stars as ambassadors. That
retainer can go even higher when the company needs the celebrity to,
ahem, throw consumers and the media off from some environmental
or social scandal. In a revelation that should shock no one, celebrities
also help companies greenwash.

Clutch your pearls, take a deep breath, and keep moving.

Why waste time? Let's go straight to the top of the celebrity
food chain and talk about everyone's favourite American family, the
Kardashians. We've seen every sordid detail of their lives, from the
moment OJ took a joy ride down the 91 Freeway to the unravelling of
Kim's ex-husband. The world expects gossip, drama, and ridiculousness
from a group of people so out of touch with reality that they think
their fortunes are self-made. Every now and then, they really throw
gawkers for a loop.

Like in 2022 when Kourtney Kardashian-Barker announced she
would be the sustainability ambassador for British fast fashion giant,
Boohoo. If you skipped over the chapter all about fast fashion, now
would be a good opportunity to go back and read it. While the industry
isn't as bad as oil, for example, that's really splitting hairs. The entire
model of companies like H&M, Zara, and Boohoo relies on more,
cheaper. Style be damned. Workers be damned. Earth be damned.

Which is why it's so odd these brands call attention to themselves and their ESG track records. Some make a big deal having high-profile sustainability ambassador roles. Boohoo isn't alone. Game of Thrones actress Maisie Williams represents H&M and British television personality Laura Whitmore represents Primark. Others, like PrettyLittleThing, set up re-sell platforms so pieces don't end up in landfill. Shein, the newest unsustainable kid on the block, announced a US$50 million fund to address textile waste.

All of these pale in comparison to a collab with a Kardashian. Maybe Boohoo's timing had something to do with a Government body naming Boohoo one of the UK's least sustainable fashion brands.[29] Or was it the reports of exploitation and unsafe working conditions throughout their supply chain?[30] Perhaps the 29 percent year-on-year increase in carbon emissions had something to do with it.[31] My guess it was all of the above. Classic misdirection layered within buzzwords about saving the planet.

The funny thing about this was the public reaction. Where you'd typically expect people to fawn over anything a Kardashian does, the public knew something was amiss. Given what I do for a living, many of my friends were very proud to tell me how they knew it was greenwashing straightaway. Media was onto it, too. Some asked how a fast fashion brand could have a sustainability ambassador. Others said this was so brazenly transparent as a cash grab it didn't even count as greenwashing. Even *Elle Magazine* called out the issue by asking, 'What Kourtney Kardashian-Barker's Boohoo Collab Says About Greenwashing'.

Not so funny, and slightly offensive actually, was just how little either Kourtney or Boohoo tried. In a mini-documentary not dissimilar to the whole Ed Sheeran fiasco, we follow Kourtney's sustainability journey. She talks to experts to inspire change (wait, is she advising the experts?), and about stuff 'like, um, worker welfare.'[32] Boohoo's head of sustainability comes on screen to 'kind of acknowledge' their environmental impact. Then, their head of corporate affairs ponders the positive effects if even 1 percent of Kourtney's followers changed their behaviour. Note it's up to *you* to change—Boohoo's fine. Kourtney then posts a really useful Instagram pic,[33] with the caption 'love her and take care of her every day' and some Earth emojis. I can hear the Valley accent as I read it.

Overall, it's breathtaking in its arrogance. I guess that matches up pretty well with the Kardashians themselves. Maybe Boohoo did make the right choice after all.

Returning to our ultimate question: does all of this good offset all of the bad? Does Kourtney's work raising awareness about sustainable fashion offset her sister taking a jet up the street? Does Joaquin wearing an expensive tux three whole times negate the millions of tons of annual emissions coming from the film industry? Will having a tofu fillet one night account for the wasteful footprint from the junk of a VIP tote? The obvious answer is a resounding *no*.

But that doesn't mean these actions are without merit. You might be wondering why I just spent so many words implying otherwise. As a way to move the needle, inspire change, and utilise their platform, even the smallest self-serving celebrity actions help. Using them as a get-out-of-jail-free card, though, doesn't fly. Be as polluting as you want. Take your jet to get the mail. Wear that Balenciaga (okay, maybe not Balenciaga) dress once and then throw it in the back of the closet, never to be worn again. Just don't do all these things and then try to tell people you care about the planet.

Although many don't realise it, we consume celebrities just as much as products from grocery stores. Metaphorically. Not literally. Unless you're Armie Hammer. What makes me so happy is that the same consumer trends we're seeing with traditional greenwashing—higher awareness, better knowledge of the issues, and a willingness to forego brand loyalty if something doesn't align with personal values—are also happening with Hollywood consumption. More than ever before, the green sheen isn't sticking like it used to. There's a long way to go, but at least we're not just accepting these things at face value anymore.

Climate Narcissists

I don't know when it happened. It just kind of snuck up on me one morning while scrolling through LinkedIn. Now, over the past ten or so years, I've done a pretty good job at curating what I see on my feed. Even though I have close to 15,000 followers (#humblebrag), I only let through things that truly educate or inform. At least, I thought so.

Then came COP 27, and the façade I'd built fell apart.

Suddenly, I began to see picture after picture of people at the UN Climate Conference. Everything was so cookie-cutter, like an Instagram post kissing in front of the Brooklyn Bridge, holding up the Leaning Tower, or hanging off that ridiculous rock in Rio. 'Glad to be in Sharm!' 'Reporting live throughout the week at COP!' 'Excited to see friends and colleagues!' It went on and on.

Was I missing something? Did my invite get lost in the mail? Why had all these people come out of the woodwork and suddenly needed to travel to Egypt for this event? It's not like we haven't been living off Zoom and Teams for the past three years. Something just wasn't adding up.

Then I saw a post that put everything into perspective. I forget who said it (so please don't sue me if it was you!), but the quote called these people out as climate narcissists. A 'climate narcissist' is a term used to describe someone who focuses solely on *their* actions and behaviour to address climate change. It's a simplistic view that doesn't consider the more significant societal and systemic changes necessary to address the issue effectively. These individuals prioritise personal image and self-promotion over meaningful action to address the climate crisis.

In my mind, they're no better than people who stop traffic to record a viral dance or go ape shit when you walk through their frame at the gym. We've been begging people to pay attention to our message for decades. Well, that seems to have come back to haunt us in some respects. That's because it's not just the ultra-wealthy or the ultra-influential using environmental and social causes to greenwash. Now, your rank-and-file social-media influencer is doing it, too.

Just like the other two groups we discussed, me lumping everyone who posted from COP together doesn't mean they're all narcissists. Out of the 40,000 visitors, I'm sure quite a few were rightfully there. But there's a delicate balance we're trying to strike. We want to get the message out to as many people as possible, necessitating everyone who cares to become an advocate and micro-influencer. People with strong expertise and a solid platform can and should go even bigger.

Where things get lopsided, and we veer into the realm of narcissism, comes down to three things. Our favourite, *motivation*, is number one. Did these people board a polluting flight to contribute, or to go on a little holiday? Second, what is their *contribution*? When we talked about my experience in Istanbul, versus the circus that is COP, real action happens when experts are in the room. Could many of these narcissists realistically impact the shape of deliberations when they couldn't even get in the meeting halls? Third, were there *alternatives*? I watched the proceedings fine from home. In a world of video conferencing and near real-time updates, if your motivation is off and your contribution nil, why are you even there? I'll give a lot of these people the benefit of the doubt. While they may be well-intentioned, was their time perhaps better spent elsewhere?

All told, climate narcissists are engaging in a souped-up version of slacktivism—activism for slackers. This is when you feel doing the bare minimum is enough. For these narcissists, the bare minimum is snapping a pic at a conference. It's worse than Kourtney talking to experts about sustainability. At least she's doing something productive. Anyone who thinks a social media post will build a better future needs to get a grip on reality. I'm not saying go out and hang off oil tankers, but come on now.

At the end of the day, the problem isn't that people are supporting causes, raising awareness, or giving to charity. These are worthwhile moves to help shape a more sustainable future. Given their status and resources, those at the top of society should be doing more than their fair share. As we mentioned, plenty are doing so in a truly authentic way. It's when the support, barracking, and giving lack sincerity. The litmus test is whether efforts are more performative than not.

I'm no idiot, either. I recognise any support to an altruistic endeavour goes well beyond doing it because it's right. There will always be ulterior motives, even if they're barely recognisable, that go into this. Sometimes—hell, most of the time—we need to tap into this to get people acting. The old adage 'what's in it for me?' applies even when you're trying to save the planet. For corporations, it's how I frame every conversation. With individuals, they need to see the personal benefit, too. When you've got some of these influential people, though, we're seeing this all happen at scale. That amplifies any tiny morsel of selfishness, shining a brighter light on true motivations.

Which brings up the three biggest issues I have with inauthentic celebrities using sustainability as a self-motivated cover. The first is that people listen to celebrities. We idolise them and believe whatever comes out of their mouths. That means people who are genuine experts get pushed to the sidelines, if they're listened to at all. There are so many great examples of this, like the ceaseless fight between Donald Trump and Dr Fauci during the height of the pandemic. One had celebrity status, and the other did not. Joe Rogan and Andrew Tate, God help us all, are inspirations to millions of young men. Gwyneth Paltrow hawks jade vaginal eggs as part of her Goop wellness brand. These are supposed to be better than anything an idiot doctor would prescribe. What do they know, anyway?

This is why listening to experts, those on the front lines day in and day out, is critical. I'm not only talking about sustainability now, either. It would help if you made this a habit throughout your life. In his famous book, *Outliers*, Malcolm Gladwell popularised the idea that it takes 10,000 hours for someone to become an expert in a subject. While many scientists refute the claim, it still stands to reason it does take a lot of time to develop an expertise. You don't become a concert pianist overnight, nor understand the intricacies of sustainable development by watching a video tutorial. In short, trust the experts, not the actors.

Second, the entire movement loses credibility as more people see these celebrities giving lousy advice, incorrect information, or just plain embarrassing themselves on red carpets. They're so front-and-centre, talking about being ambassadors for a movement they may know very little about. Like activists, that makes them representatives to the general public. It doesn't matter how many experts work their asses off behind the scenes. All this can be easily undone by a laughingstock parading as a know-it-all.

Hollywood is notoriously focused on image. We've got to be smarter. While you're listening to experts, why not also focus on those who are true ambassadors? Angelina Jolie, Leonardo DiCaprio, and Alyssa Milano may not be perfect do-gooders (spoiler alert: there is no such thing). But they're better ambassadors and role models than most other celebrities. They've spent time learning from experts and experiencing things on the ground. Yes, Jolie and DiCaprio like their private jets and superyachts, so this isn't giving them a pass. If we're talking about positive contributions outweighing the costs, both have certainly banked a bit of leeway.

Finally, there's no accountability for their antics. Ed Sheeran could have done real damage going to Liberia, especially if he had to start selecting who he'd pay to house. Kourtney's cash grab was great for her but exposed an entirely new segment of her fans to the horrors of fast fashion. Elon is literally using his new toy to spread social, political, and environmental misinformation. But nothing much happened beyond a Rusty Radiator Award, bad press, and some viral GIFs.

This means it falls on us to make things happen. The question is how to hold their feet to the fire and make them accountable. I've thought long and hard about an answer to this question, but everything I've come up with seems trite and basic. Maybe that's because narcissists prey on the naïve, so doing the bare minimum is enough to get the momentum going. All that starts by avoiding these delusions of grandeur. Arnold may be an idol when he's kicking someone's ass on the big screen. The second he hops into his gas-guzzling Hummer, he becomes a supervillain. Taking them down off a pedestal makes it much easier to see them as fallible and call out their hypocrisy.

As the long-running *Us Weekly* article says, 'Stars—they're just like us!'

Chapter 18

You ... Yes, You

Pop quiz!

You're at the grocery store and see a display from one of those capsule coffee companies. Signage prominently shows a picture of smiling plantation farmers. The tagline and copy talk all about their amazingly sustainable production practices. There's even a leaflet you can take home with more information. What's wrong here?

... Exactly. While the production itself might be sustainable, coffee capsules are notoriously bad for the environment. They end up in landfills by the millions and take centuries to decompose.

Question two. While walking downtown with some of your friends, you run into a rowdy climate protest. Activists are shouting slogans, handing out signs, and throwing paint on buildings. Your friends are getting riled up with the crowd and are getting ready to join in. What do you tell them?

... Perfect. You've learned there are more pragmatic ways of communicating a message. Getting angry and breaking shit isn't the right approach. It's much better to have a plan and work that plan.

Final question. A sibling, friend, or parent notices you reading this book. They come up and ask, 'What is the number one thing I can do to save the world?' What's your answer?

Recycle.
Wrong.

Speak with our wallet.
Closer, but still not quite right.

Take public transport.
You're going the wrong way now.

Along our journey we've learned a lot. By now, you can spot when companies try to misdirect you with green speak. It should be easier for you to sift through an influencer's lies about supporting the environment. When you see a slick video talking about the latest eco-innovations from this country or green investment from that billionaire, you'll automatically smell if something's off. That's because we've covered greenwashing by everything from big corporations to fast fashion houses to individual governments, and everyone from the rich to the famous to the influential.

But now we come to the most important greenwasher of all: you.

Yes, even you greenwash now and then.

Personal greenwashing is less about the things we buy or the causes we support. It involves the intrinsic and preconceived notions we've inherited over time. None of that's your fault, of course. But nearing the end of this book, and having educated yourself on the subject, it is *your responsibility* to change. So, we're going to explore some of the ways personal greenwashing might manifest in our daily lives. That includes some of the big falsehoods and misconceptions you should know about when it comes to saving the planet. We'll also dive into how educating yourself may seem like a good idea, but can sometimes lead you astray.

By the end of this, you'll be able to confidently explain why saying yes to everything is killing the planet, what recycling really accomplishes, and how condoms might just save us all.

P.S. If you want to find out the answer to that third question, keep reading.

Manifest This

Just like there are no perfect corporations, the perfect environmentalist doesn't exist either. Even Greta, Al Gore, and the looniest activists do things we might not consider eco-friendly. And that's okay! This notion that to be doing good means you can never do something 'bad' is a false dichotomy. It's a way for people who don't care about the Earth to point at us that do and say, 'Ah, hah! You call yourself an environmentalist and you ate a hamburger/drove a car/watered your lawn.' Then they'll use this ridiculous sentiment as an excuse to sit back and watch the world burn.

Where did we get this idea? I track it back to those carbon footprint calculators you hear people quoting. Fun fact. These calculators are an exercise in greenwashing themselves. In the early 2000s, oil giant BP hired public relations firm Ogilvy to fix its image. They came up with the term carbon footprint, a genius way to pass the blame for pollution onto consumers. In 2004, BP went one step further and launched the first carbon footprint calculator. Today, everyone loves to check how small (or large) their footprint is. Little do they know it's all funded by a fossil fuel behemoth.

But here's the thing. These calculators aren't representative of your whole life. Some of the typical questions on a calculator ask how often you fly, drive, or walk, whether you're a vegan or eat meat, and if you live in a city or the country. Maybe this gives an indicative level of emissions, but how about other ways you might contribute to a better future? If I look at my life, the calculator would give me a fairly large footprint because I have to travel for work. Work, mind you, that's explicitly involved in saving the planet. Do I get a carbon footprint concession? Do you get credit for recycling, giving to a non-profit, or planting a garden? No, because it's nothing more than a fictional measurement.

Which brings us back to the idea of perfection. We live in a modern world with lots of distractions. Our lives are multi-faceted and complex. Even for the most die-hard environmentalist, saving the planet will only be one part of who they are as a person. Unlike the carbon footprint calculator, you have to give yourself credit where it's due, forgiveness where it's necessary, and the chance to enjoy the Earth you're helping to protect. Trying to do everything will accomplish nothing. Struggling to reach the unattainable goal of perfection will lead to burnout. You're delusional if you think otherwise.

When we look at our personal sustainability journeys, the goal isn't perfection. If it were, we'd all be greenwashing by missing the mark. The goal is to improve. As long as there's the right balance in your life and you're moving in the right direction, I'd count that a win for the Earth. In the spirit of improvement, a great starting point is looking at how greenwashing might manifest in our daily lives. Again, this has nothing to do with things like purchase habits. Instead, it's how we think about, respond to, and act on building a more sustainable future.

First up, is pure ignorance causing you to greenwash? If it does, I'd put you in a group I call the *viridi stultus*. Latin for the green fool, these are people who innocently greenwash because they don't know any better. Maybe they're only starting their sustainability journey, unaware of the options available. That means their efforts might be misplaced. Some may have fallen into the trappings of the vast amounts of misinformation that's accessible. Misguided, this could lead them to believe something that isn't precisely correct, like that recycling is all it takes to save the planet. Perhaps you're an eco-warrior from way back but are stuck in outdated ways of doing things.

The world of sustainability changes so quickly it's often hard to keep up, especially recently. But by not keeping up with progress, you've wasted time digressing. Honestly, I think we all have a bit of the green fool in us. There's always something more we can learn, try, or do. The key is to stay informed and vigilant to guard against this accidental greenwashing.

Maybe you're up-to-date with all the ways you can be more sustainable. You've spent lots of time fighting the good fight. Now, you're tired and have started to become a bit of a slacktivist. Do you find yourself thinking, *do as I say, not as I do*, more often than before? Is saving the planet a young person's game? Are you scrolling through social media and considering giving up since we're screwed anyway? Sorry to say, but too much is given, much more is required. If you are an elder in our tribe, you must work twice as hard given your knowledge and experience. There's no room for performative actions or letting all that knowledge give you a superiority complex. From one elder to another, we've worked too hard to get where we are. We've got the wind at our backs, so it won't be nearly as difficult as in the past. Now is the time to kick things into overdrive and really positively impact the world. Anything less is just greenwashing.

So you know what there is to know, have the energy, and are genuine in your intentions. This is probably the majority of you reading. Be careful, though, to avoid becoming a micro-manager. I'm not talking about that annoying boss always hovering around your cubicle. A sustainability micro-manager is someone who's become an expert at compartmentalising their good deeds. They do this to bank up credits, knowingly or not, that offset some of the less sustainable things going on in their lives. Imagine a vegan who drives a big SUV or someone who recycles and composts all the waste from their massive mansion. It could be an oil activist who smokes three packs a day or the bitcoin bro who screams at you to save electricity.

Psychoanalyst Diane Coutu talks about this segment feeling very conflicted, and I could see why they would. I'd argue that combatting these feelings only requires asking a few critical questions. First, are you deluding yourself? Smoking is just as harmful to the environment and humans as oil production. Bitcoin mining takes up a massive amount of energy. Offsetting that will take more than turning off your bedroom lights. What you're doing in one aspect of your life isn't truly counterbalancing some of the other stuff you're engaging in, is it?

It genuinely is, and you're still feeling conflicted? Then ask yourself if there's too much self-imposed pressure to be perfect. Remember, this isn't the goal. Instead, be honest and transparent about your actions and actual impact on the planet. Understand you are only one person. Remind yourself you're doing all you can. And if you want to take that flight to Bali, go for it. Don't forget to strike a balance when you get back.

Little Old Miss Conception

Mark Twain famously wrote, 'The trouble with the world is not that people know too little; it's that they know so many things that just aren't so.' I couldn't agree more, especially when it comes to sustainability. We've picked up some pretty hefty baggage over the past century of trying to save the planet. Like many of the approaches activists and NGOs use today, our baggage served a purpose at one time or another. Today, though, all it's doing is unnecessarily weighing us down.

What is this baggage? It's the misconceptions we have about what it'll take to build a better future. It's the preconceived ideas about right and wrong. It's the tunnel vision and inherent biases that prevent us from innovating and learning. Boiled down, these are all distractions. Calling them out is important because they're preventing us from getting the job done.

Throughout our journey we've already encountered quite a few. I've harped on and on about how we need to rely more on the private sector to do the lion's share of the work, why we should be converting more than raising awareness, and, most recently, why there's no such thing as a perfect environmentalist. These ideas run counter to the prevailing notions that activists, activism, and some sort of magical switch are all it takes to move forward. We've learned why groups like the United Nations, and influential meetings like those of the World Economic Forum in Davos, aren't all they're cracked up to be. I've based this whole book on the idea companies want you to have one perception—a misperception—of what they are.

These misconceptions permeate our daily lives, as well. They impact our thinking and, in turn, our actions.

Changing thought patterns is easier said than done. As altruists, the biggest one we have to change is this notion we're in it alone. I'll let you in on a little secret. You're not the only person in this world who cares, and that's a great thing. There is an army of fellow altruists out there ready to make a positive change in some way, shape, or form. Just crunch some back-of-the-napkin numbers. If there are around eight billion people in the world, of which a quarter are under fifteen years old, that gives us a total population of six billion potential do-gooders. For the sake of argument, let's be super pessimistic and say only 10 percent of the world's people care about things beyond themselves (it's obviously much higher). That still gives us 600,000,000 altruists around the world. Not a rag-tag militia, if you ask me.

Which leads to the second big thought pattern we need to break: saying yes to everything. When it comes to saving the world, this predilection towards saying yes hasn't done us any favours. You struggle to not spill the ethically sourced latte out of your reusable thermos while riding a bike to the animal shelter. Your small electric car is so filled with banners and leaflets there's no room to pick up trash from the side of the highway. You want to read to the elderly, feed the homeless, and help raise money for a local campaign. But you're only one person with a finite amount of time, capacity, and sanity. How do you expect to get all this done?

Remember—you can do anything, but you can't do everything.

That's why it's critical to prioritise. One of the best prioritisation methods is the Eisenhower Decision Matrix. Named after the former US president, it's an effective way to decide the criticality of a situation. As Eisenhower often said, '... what is important is seldom urgent and what is urgent is seldom important.'[1] To prioritise his time as Supreme Commander of the Allied Expeditionary Forces in Europe during World War Two, Eisenhower would put things into four quadrants: important and urgent; important, but not urgent; not important, but urgent; and not important, not urgent. In the 'important and urgent' quadrant would go things like crises, deadlines, and ongoing problems. Important, but not urgent, issues included relationships, planning, and recreational activities. Further down in importance were things like meetings. Then there were trivial things—time-wasters—relegated to the 'not important, not urgent' quadrant of the matrix. It helped him win a war. It can help you figure out what's most important to saving the world.

Even the best prioritising is only useful when you follow through. You can't put your plan together then fall back into the habit of saying yes to everything. It's taken me years to become laser focused on my part of the sustainability universe. I've trained myself to ignore the people who say I'm heartless for not giving to this charity or supporting that cause. Muscle memory now stops me from taking every starving dog home because I know I can't save them all. It's tough turning my back, but I know I have limited capacity that needs to be focused elsewhere. In essence, I learned how to say no to a lot of things.

The best-selling 1981 book, *Getting to Yes*, discusses six essential elements of successful negotiation.[2] As a former negotiator, I would refer to this book often. Counterintuitively, some of the six tactics in *Getting to Yes* are useful in figuring out how to say no. One is to step back and separate the people from the problem. Taking a vested, personal interest in something makes it appear much more important. Having a critical, impartial view of a situation can help you decide whether it's worthy of your time. This viewpoint will also help you manage emotions. We often say yes instinctively; a gut reaction seldom based on logic. Getting to yes or no has to be more deliberate. In saying no, you'll disappoint people. They'll try to change your mind. Instead of caving, put a positive spin on your choice. Explain your laser focus on areas best matching your skill set. Talk about the other great organisations already assisting X, Y, or Z cause. Decision-making, free of emotion and full of pragmatism, will work to maximise your impact.

With your army behind you, and pragmatic prioritisation in place, it's time to supercharge that impact. You can do this by shifting your viewpoint just a touch. These shifts were *ah-ha* moments for me, shaping a formative change in my approach to saving the planet. Number one is to get rid of this idea we're trying to save some future generation. Activists and politicians have been using this line for decades. Look at what's happening around you, and you'll realise we are that future generation. The work we're doing today is to save ourselves. Hopefully, that makes it all hit a bit closer to home.

It's also high time we throw out the patronising belief that western countries have all the answers. West isn't necessarily best. I often half-jokingly tell my colleagues in the US and EU they view sustainability as an academic exercise. Meanwhile, those of us working in the developing world have to roll up our sleeves and get our hands dirty. Having spent a decade in China, where sustainability issues are constantly in your face, I know the solutions are happening apace. China invests double into green technology every year than the entire European Union while also having a long list of superlatives to add to its resume—the world's biggest electric vehicle market; the world's longest high-speed rail network; the world's largest solar and wind energy producer. The list goes on. Maybe there's a thing or two we can learn if we get out of our nationalistic bubbles.

Finally, we have to be hyper-vigilant against the rising tide of climate doomism. Clearly the world should be in a much better state than it is currently, but that doesn't mean we're too late to change things. Many in the media would like you to believe otherwise. Author Johnathan Franzen claims, '... [w]e literally are living in end times for civilisation as we know it ... We are long past the point of averting climate catastrophe.'[3] Influential TikToker Charles McBryde confessed to being a doomer: 'Since about 2019, I have believed that there's little to nothing that we can do to actually reverse climate change on a global scale.'[4] Just like every messianic end-times cult, they couldn't be more wrong.

A significant increase in access to information is vital to understanding the rise of doomism. In our highly polarised world, where we frown on civil debate and listening to alternative viewpoints, if it bleeds it reads. Tragedy, doom, and gloom get the clicks. In no way am I wearing rose-coloured glasses. As someone who deals with this stuff on a daily basis, I'm well aware of our state of affairs. There are great things, positive things, happening too. Only giving airtime to the bad stuff leaves little room for balance. This one-sided focus has provided a platform for self-interested parties posing as experts to scare the bejesus out of people.

It's also given greenscamming groups, especially from the oil and gas industry, the opportunity to infiltrate the conversation. Along with a rise in doomism has been a rise in climate misinformation. A 2022 report by The Institute for Strategic Dialogue (ISD), Friends of the Earth, The University of Exeter Seda Lab, and the Union of Concerned Scientists[5] noted a stark comeback of climate misinformation and denial on social media. The report '... identified fossil fuel-linked entities that spent about $4 million on Facebook and Instagram ads around the time of the United Nations' climate change conference in November.' I've seen the viral hashtag #climatescam come up more than a few times as a top search term on Twitter. Looking through the threads, highly questionable stats, graphs, and reports all point to a lot of BS making its way into our heads.

Several global research teams have dug specifically into climate misinformation on Facebook.[6] One group found '78% of the [climate misinformation] advertising spending identified came from just seven pages'. At the same time, another noted that 'just 10 publishers are responsible for 69% of digital climate change denial content on Facebook'. People are looking at this stuff. A third study condemned Facebook as a bastion of distortion, reporting twenty-five million views of climate-related misinformation over a short two-month period.

Like its little cousin eco-anxiety, climate doomism is a fear tactic that paralyses. It's also a foundational pillar for the eco-slacktivist. I'd like to tell you not to fall for it, but doomism is far too sophisticated a beast. Instead, balance and moderate the information you consume. If that doesn't work, remember who is best served by all this doomism: greenwashers. The fossil fuel companies, corrupt politicians, and those stuck in the past are thrilled with us doing nothing. In the words of Brynn O'Brien, executive director of the Australasian Centre for Corporate Responsibility, '... [t]he only people who fall for [climate doomism] are rich white people who think they will be spared until everyone and everything else is gone.'[7] ·

Changing your thinking is the hard part. Adjusting some of the long-held beliefs about what it'll take to improve the world should be a piece of cake. It's probably clear by now I don't ascribe to this idea we have to return to the Dark Ages to save the planet. Giving up all our modern conveniences—international travel, mobile phones, cars— ain't gonna happen. Some people might want to go live off the grid. More power to them. By and large, though, most of us like our creature comforts.

Another sacrosanct belief we have is in recycling. We assume it's foundational to 'doing our part'. I'm from Southern California, so it's kind of in my blood. Recycling is so established I doubt you even question whether it's having an impact. As far as the stuff you and I are throwing away, the answer's no. That's because less than 10 percent of all recycling waste comes from normal households. The other 90-plus percent is thanks to big business. When we look at plastic pollution in particular, only about 9 percent of all plastic ever gets recycled. Most of it ends up in the landfill, leeching its toxins into the ground for the next thousand years. I'm not telling you to stop recycling. Don't deceive yourself, though, in thinking that recycling's good enough.

How about the movement to get rid of plastic straws? It seemed, for a time, the oceans must have been full of them. In fact, plastic straws make up only 0.025 percent of all plastic ocean waste.[8] This isn't to say the cause wasn't laudable or successful. To be clear, it was both. My point is we could have spent all that time, effort, and ad money focusing on the other 99.975 percent of plastic waste ending up in our precious waters. It's like trying to fix a leaky faucet when there's a tidal wave bearing down. Let's focus on the big, important stuff first.

Consider, too, the topic of charitable giving. Since the late 1970s, old white guys have come on television asking us to sponsor a child for only two dollars a day. That's less than the price of a cup of coffee (not in Melbourne!). Lots of people give, but where's the money actually going? World Vision, one of the biggest humanitarian organisations utilising TV advertising, promised money would go towards a number of different things that could make a difference, such as clothing or education. In 2008, an Australian reporter visited a teenage child he had been sponsoring in Ethiopia. Besides a World Vision pen, the girl knew nothing about the sponsorship. Although the reporter received progress reports, especially on how much the child's English was improving, in person she spoke no English at all. It seems the cheque got lost in the mail.

Finally, we come to the biggest question of all: what is the number one thing we can do to ensure a more sustainable future? Is it shutting down factories that spew out polluting greenhouse gas emissions? How about cleaning up the massive bergs of plastic littering our oceans by reducing our consumption? Or could it be halting the highly polluting agriculture sector by going vegan? While all these are absolutely things we should be doing, none are the correct answer.

That's because the number one thing all of us should do if we want to save the planet is to have fewer children. According to IOP Science, having one fewer child will reduce annual personal emissions by an average of 58.6 tons of CO_2 if you live in the developed world.[9] Compare this to living without a car, which only reduces your emissions by 2.4 tons. Going on a trendy plant-based diet will only reduce your emissions by a measly 0.8 tons a year. It gets worse. Recycling, which we all hold dear, reduces annual emissions by less than 0.1 tons. That means having one less child is exponentially more impactful than most of the things we're doing today to combat climate change.

You might need a minute to process all this. I just upended a lot of what we've held sacred for a very long time. None of this is to dissuade you from continuing all the amazing things you've been doing. To be clear, recycling, giving to charity, and reducing consumption are invaluable contributions. I want to emphasise the importance of thinking critically about everything we do. That way, we can maximise our positive impact on the planet instead of doing things out of habit.

Avoiding greenwashing in your personal life takes having a pragmatic eye pointed at everything you do. In my first book, *Sustainability for the Rest of Us,* I lay out five foolproof ways anyone can become more pragmatic and save the planet in the process. They're so salient that applying them to personal greenwashing works just as well. Consider this the very abridged version of that book (but don't think this gets you out of buying a copy and reading the whole thing!).

Firstly, you have to educate yourself. Reading *this* book is a great step toward getting up-to-speed with the latest sustainability thinking. There's lots more to learn, so consider resting on your laurels taboo. Learn the fundamentals of sustainability so you can make an informed choice. Be cautious of misdirection in your own life and from others. Connect with experts, read voraciously, and perfect that Twitter feed. Most importantly, stay up to date. Now that your eyes are open, the last thing you'd want is to return to the world of the *viridi stultus*.

Second, remember you can do anything, but you can't do everything. You've got to focus and contribute but not get to the point of burning out. As we saw with the slacktivists, it's very easy to get bogged down in the work and forget the purpose. Even though we're pushing shit uphill, it's to build a better future for humanity. To keep focus, find your passion points. They could be something you enjoy, have expertise in, or are financially able to contribute to. Passions are passions because you're passionate about them. If you've lost that spark, that flare, find something that reignites it. Keep the fire burning bright!

Next, assess your lifestyle. Take an honest look at your daily habits, routines, and attitudes. Identify areas where you can make changes to reduce your impact on the environment, increase your positive impact on society, or build that army of altruists. Don't forget about those all-important influences on your life, too. It might not necessarily be someone sitting in an ivory tower. Consider the influences in your friend groups, at work, and even in your family. If you want to become a battle-hardened sustainability pro, it's up to you to gather the right people around you.

Then, set out your plan of attack. Personal greenwashing is more likely to happen when we shoot from the hip. A plan guides our work and shows us where we can improve. It also guards against unintended consequences as well as a mismatch between words and actions. Firstly, take stock of all the things you're already doing. Then, use your plan to keep track of your progress and hold yourself accountable. Be transparent with your efforts, acknowledge your failings, and always celebrate your successes.

Lastly, get out there and get to work. Dipping your toe in and out of the proverbial water doesn't do anyone any favours. In fact, it often does more harm than good. Once you know what it is you want to do, and can do, then go out and jump in with both feet. Don't half-ass anything. There is no space left for trying this whole sustainability thing out. We're really up against the clock, so if you want to make a difference, don't wait for the right time. You'll always find an excuse for why it isn't. Do it now.

In his book *Atomic Habits*, James Clear talks about how small habits can make a big difference. He mentions the Sorites Paradox, an ancient Greek parable on the importance of cause and effect. Clear also alludes to the CBE improvement program by the Los Angeles Lakers basketball team and a similar system set up by the British National Cycling Team. The author asks what impact one small action can have when repeated over a long period of time. Well, it doesn't matter if you're Greek, American, or British, small actions can have a tremendous impact.

> It is so easy to overestimate the importance of one defining moment and underestimate the value of making small improvements on a daily basis. Too often, we convince ourselves that massive success requires massive action.
>
> Meanwhile, improving by 1 percent [sic] isn't particularly notable – sometimes it isn't even *noticeable*—but it can be far more meaningful, especially in the long run … if you can get 1 percent better each day for one year, you'll end up thirty-seven time better by the time you're done.[10]

Dear reader, this is the message I want to leave with you. When it comes to being the most sustainable version of yourself, remember to think less in perfection and more in small, consistent changes. If it worked for Magic Johnson, it'll work for you too.

Chapter 19

Build Your Army

There's a mantra that business leaders, wellness gurus, and celebrities like to repeat. Attributed to motivational speaker Jim Rohn, it claims you're the average of the five people you spend the most time with. Put another way, you are the company you keep. If you want to be a more successful businessperson, as the mantra goes, surround yourself with successful businesspeople. Want celebrity? You better get yourself to Hollywood. Money and power? Cosy up to the person in the corner suite. Like the cliques in an American high school cafeteria, where we sit determines who we are, how we're perceived, and where we're headed in life.

Okay, well I wouldn't go so far as to say those you hung out with in high school pre-determined the rest of your existence. If that were the case, I'd still be a dorky musician. But our influences—both known and unknown to us—shape our lives in more ways than we realise. The same holds true with how we attempt to save the planet and our future. Although one might assume they are surrounding themselves with the right people and messages, hyper-vigilance must be the watchword.

Throughout this section we've explored some of the main influences, both past and present, impacting our mission of building a better tomorrow. Poster children of the movement, those in the activist and NGO communities, are the most influential. While I ardently support these groups, some of their approaches must change. Activists lack pragmatism. Moralists lack empathy. NGOs have sold out. If they're going to continue to be on the frontlines, then they need to be better at getting the message out there. I'd argue they need to begin thinking less like activists and more like marketers. That means having a stronger understanding of who their target audience is and what they want to hear. If not, these groups risk blocking new potential support, haemorrhaging the supporters they already have, and making our already difficult job even harder.

Next, we turned to the ultra-wealthy, ultra-famous, and ultra-influential. While their work supporting causes, raising awareness, and giving to charity should be part of their responsibility as the upper echelon of society, it's also fodder for the media. Thus, ulterior motivations will always play a part. Yet there's a big difference between those who give graciously and with purpose versus those inauthentic celebrities using sustainability as a self-motivated cover. With them, I have three big issues. The first is in the influence they wield and how the public idolises their every word. That sidelines real experts doing the hard work. Second, giving unfounded advice and touting performative eco-lifestyles causes credibility issues for the movement. Lastly, there is a general lack of accountability for their actions. Unlike Gal Gadot's ridiculously out-of-touch *Imagine* disaster, which we laughed at and forgot, with the Earth ... the stakes are much higher.

Finally, we took a deep look at ourselves and some of the ways greenwashing might show up in our personal lives. Sometimes this can happen inadvertently and other times through laziness. We may also compartmentalise our good to offset our bad. No matter how it manifests, we can combat personal greenwashing with a slight shift in our thoughts and deeds. That means remembering you're not in this fight alone. Millions of people, just like you, want to build a brighter future. It also means you've got to prioritise all the good you're doing. You'll risk burning out in trying to do too much, and nobody wants to see that! Be vigilant against doomism and misinformation, consider whether your actions are accomplishing what you think they are, and don't worry about being perfect. There's no such thing!

When it comes to the intersection of influence and saving the world, there are three key things I want all of you to remember.

First: greenwashing, like influence, can occur in many forms. Think about all the different sources we've discovered. Corporations greenwash in a mind-boggling number of ways. Politicians greenwash, whether at the local, national, or international level. Your favourite sports teams, celebrities, and charities do it, too. There are so many layers of complexity. Sometimes, that makes it very difficult to discern when it's happening or where it's coming from. Staving off greenwashing requires the analytical skills of an NSA agent and persistence of an investigative journalist. All this work is why trust should be earned, not given lightly.

Second: influence, like greenwashing, is often self-motivated. Whenever money, clout, or personal advancement enter the chat, you must take a step back and think critically about the all-important why. What's the motivation behind why a group is doing something? Is there more to a story than what's on the surface? Where does this idea, practice, or action come from? In an age of increasing misinformation, keeping on one's toes is more important than ever. Be sure to vet any and all who would try to influence you. Think for yourself instead of letting others think on your behalf.

Third: nothing is sacred. The companies you buy from, causes you support, and influencers you admire can all change. How you think about saving the planet and what you do to help aren't set in stone. For the past century, we've held onto certain beliefs as if they were inviolable. In many ways, this has prevented us from having actual progress. Instead, we must all employ a greater pragmatic view to ensure the most significant impact possible. A more philosophical way of putting this would be to say our destination is fixed, but our pathway may change. In business, it would be to fail forward but keep going.

Keep going.

A Future Worth Fighting For

Remember way back when you started reading this book? I asked you to imagine yourself standing at the supermarket shelf. You were looking at the rows and rows of products, trying to figure out which coffee or tea to buy. It probably seemed a pretty easy decision at the time. Your mind would have gone through a few quick scenarios, weighing out the costs versus benefits to land on your final choice.

If we ran that experiment now, you'd likely have a much harder time. That's because I've gone and filled your head with a bunch of new ideas. More specifically, I've given you many more questions to ask yourself whenever you buy anything. But because of this, you're now able to make a more informed decision that can, in turn, have a real positive impact. That impact can be at the grocery store or the high street, curating your social media feed or group of friends, in how you think and what you do. Information is power.

Over the course of 300-ish pages, you've had a lot of that power thrown at you.

We began by defining greenwashing, what it is, and why it has become so pervasive. You learned about its evolution from a small boutique hotel in Fiji to the massive oil cover-ups still going on today. The perfect storm of endless new products, technological advancement, and the consumer need to do good has become the hallmark of today's greenwashing. That greenwashing comes in three primary forms: green speak; misdirection; and greenscamming. With each, we talked about how to spot the lies they're peddling and what to do when they try one on you.

Next, we pulled back the curtain on greenwashing in the corporate sector. Our starting point was getting the truly unsustainable industries, like tobacco, defence, and oil, out of the way. Since these industries can never be sustainable (unless they put themselves out of business), there's no point in ever believing any green claims they might make. From there, we explored the range of industries that are bad today but have the space to change. Among those were aviation, automotive, and some fast-moving consumer goods. Our discussion also highlighted one of the most prominent poster children of greenwashing today— the fast fashion industry. Ultimately, I explained my firm belief in the private sector being the most significant driver for positive change.

Moving on from the consumer sector, we set our sights on public sector actors and institutions. Although this is the one group we assume should have the planet's best interests at heart, they far too often fall short of expectations. It's very easy to see right through their veneer of empty platitudes and unachieved targets. From the United Nations to individual countries to global organisations like the International Olympic Committee, each has made a habit of greenwashing, not unlike the big corporate players we like to vilify.

Then it came time to discuss those groups and individuals most influential in our lives. For years, many of them have put us on the wrong trajectory to save the planet. I outlined three big groups doing so. Activists, moralists, and non-governmental organisations—while they might mean well—have become hamstrung by their lack of pragmatism. The ultra-wealthy, ultra-famous, and ultra-influential are highly admired but often very misguided. Finally, we put a mirror up to ourselves and our actions to see what we could improve.

All of this has probably left you with two very big questions: why should you care; and what should you do?

By picking up this book, you've already demonstrated why you should care. That's because the future of the planet depends on it. Beyond this, though, understanding greenwashing has more tangible impacts. Understanding greenwashing puts the proverbial ball back in your court. You can now wrest control back from corporations, organisations, and individuals who would purposely lie to you. No longer can they assume your loyalty at face value. Now, they must earn it. In the end, this helps prevent you from being rolled up as an accomplice to factions who value profit over the planet.

So, you care. Great. But what in the world are you supposed to do in a few short moments when you close the back cover of this book? Firstly, don't just put it back on the shelf to collect dust. Think of *The Great Greenwashing* as a manual to come back to again and again. Then, consider the reasons why you chose to read *The Great Greenwashing* in the first place.

If you picked up this book because you wanted to leave a better world for your children, I'm guessing you already knew a little something about greenwashing. You may have heard about it on television or seen the word in the newspaper. Now you're informed and able to influence the future positively. Maybe you bought this book out of sheer curiosity. What the hell is this guy John going on about greenwashing for? You, too, have had new canyons of information etched into your brain. My challenge for both of you is to take all of this new material and use it in your daily lives. It's definitely easier said than done. But take it step by step and eventually you'll find that filtering through the greenwashing nonsense has become second nature.

For those deep in the trenches of the sustainability profession, much of this book may have been a good refresher. When we're head down in the day-to-day slog of saving the planet, it's easy to lose the forest for the trees. That's why it's so important to occasionally take a step back and get out of our echo chambers. Reading the opinions of others, as you've just done, is an excellent way of accomplishing that. Now, one of three things can happen. You can have your opinions validated or thoughts vindicated by the words on a page. You might also discover things you disagree with, reinforcing your existing set of beliefs. There's also the third scenario, where you learn something new and have your worldview shaken, moulded, and expanded. Maybe this book was a combination of them all.

Finally, some of you may have initially been excited to read all about greenwashing. Page by page, though, you've realised it's you I've been writing about. First off, kudos for having the level of introspection required to reach that point. But, not to worry! Just like Saint Paul on the road to Damascus, there are always opportunities to change. This is your wake-up call and challenge to break old patterns of behaviour. For readers from the corporate world, that means taking a leadership position to enact change in your organisations. People toiling away in government or international bodies, it's time to assess whether your work really has the intended impact you want people to remember you for. And for those in a position of influence (hi, Kylie!), now is when you do your *mea culpa* and start to use your platform for good.

I'm not a futurist, but people often ask me where I see sustainability in ten or twenty years. As I've said over and over, my strong belief is that the same corporations I've torn apart in this book will bring about a more sustainable future. By the time I'm fifty, I want to be confident that every purchase I make has a positive impact. Through innovation and the pure capitalist desire to survive, the private sector has already entered this virtuous cycle. Savvy organisations are developing increasingly better, more sustainable products. Over time, those hold-out companies that refuse to collaborate will find it harder and harder to get a seat at the table. Eventually, they'll wither and die.

Like with modern technology, the rate of change will then become exponential. In this future, private sector companies will have already done 95 percent of the work. By the time you get to the supermarket shelf, every single option will be a sustainable one.

This is the future I'm looking forward to. No matter how you found your way to this book, I want you to look forward to the future, too. It's never too late to make a positive impact. With all the doom and gloom out there, I know this might seem like a tall order. But staying focused and hopeful is what will speed us towards this better tomorrow for you, your loved ones, and the planet.

Endnotes

PART 1: GREENWASHING: A CRASH COURSE

Chapter 1: Did You Hang Up Your Towel?

1   "Greenwashing," *Investopedia*, accessed 1 March 2022, https://www.investopedia.com/terms/g/greenwashing.asp.

2   Alex Carey and Andrew Lohrey, "Taking the Risk Out of Democracy: Corporate Propaganda Versus Freedom and Liberty." *University of Illinois Press*, 1997, https://www.azquotes.com/quote/529529.

3   David Leonhardt, "The Charts That Show Big Business is Winning." *The New York Times*, 17 June 2018, https://www.nytimes.com/2018/06/17/opinion/big-business-mergers.html.

4   **Corporate mergers:** "History of the RJR Nabisco Takeover." *The New York Times Archive*, 2 December 1988, https://www.nytimes.com/1988/12/02/business/history-of-the-rjr-nabisco-takeover.html; Klaus Ulrich, "Mannesmann: The Mother of All Takeovers." *Deutsche Welle*, 3 February 2010, https://www.dw.com/en/mannesmann-the-mother-of-all-takeovers/ a-5206028.

5   Harry Bradford, "These 10 Companies Control Enormous Number Of Consumer Brands [GRAPHIC]." *The Huffington Post*, 7 December 2017, https://www.huffingtonpost.com.au/entry/consumer-brands-owned-ten- companies-graphic_n_1458812?ri18n=true.

6   "Cendant Corporation." *CFA*

Institute, accessed 28 April 2022, https://www.econcrises.org/2016/11/29/cendant-corporation/.

7   **On water use in hotels**: "Hotel Water Use: Are You Flushing Money Down the Drain?" *Environment and Energy Leader*, 21 July 2016, https://www.environmentalleader.com/2016/07/hotel-water-use-are-you- flushing-money-down-the-drain/;
"A Truly Clean Matter – Steps Toward Eco- Friendly Laundry in Hotels." *Green Pearls*, 24 January 2019, https://www.greenpearls.com/newsroom/a-truly-clean-matter-steps-toward- eco-friendly-laundry-in-hotels/.

8   "Implementing a Linen and Towel Reuse Program." *Guest Supply*, 18 December 2019, http://thesolutionsdesk.com/implementing-a-linen-and-towel-reuse-program/.

9   **On corporate greenwashing**: Bruce Watson, "The Troubling Evolution of Corporate Greenwashing." *The Guardian*, 20 August 2016, https://www.theguardian.com/sustainable-business/2016/aug/20/greenwashing-environmentalism-lies-companies;
"Greenwashing Costing Walmart $1 Million." *Environment and Energy Leader*, 3 February 2017, https://www.environmentalleader.com/2017/02/greenwashing-costing-walmart-1-million;
"Everything You Need to Know About the VW Diesel-Emissions Scandal." *Car and Driver*, 4 December 2019, https://www.caranddriver.com/news/a15339250/everything-you-need-to-know-about-the-vw-diesel-emissions-scandal/.

10   "More Than Values: The Value-Based Sustainability Reporting That Investors Want." *McKinsey*, 7 August 2019, https://www.mckinsey.com/business-functions/sustainability/our-insights/more-than-values-the-value-based-sustainability-reporting-that-investors-want.

11   "How Many Products Are In A Typical Grocery Store?" *ICSID*, accessed 1 June 2022, https://www.icsid.org/uncategorized/how-many-products-are-in-a-typical-grocery-store/.

Chapter 2: How to Spot Greenwashing

1   Ibid., *Investopedia*.

2   "What is World Without Waste?" *The Coca-Cola Company*, accessed 17 June 2022, https://www.coca-colacompany.com/faqs/what-is-world-without-waste.

3   Ibid., *Investopedia*.

4   Ibid., Watson.

5   "EcoLabel Index." EcoLabel, accessed 2 June 2022, https://www.ecolabelindex.com/.

6   "Destruction: Certified." *Greenpeace*, 10 March 2021, https://www.greenpeace.org/international/publication/46812/destruction-certified/.

7   "Greenwashing Tactic #4: Fake Labels." *Palm Oil*

Detectives, 22 October 2021, https://palmoildetectives. com/2021/10/22/greenwashing-tactic-4-fake-labels.

8   Ibid., *Greenpeace.*

9   Christina Caron, "Starbucks to Stop Using Disposable Plastic Straws by 2020." *The New York Times,* 9 July 2018, https://www.nytimes.com/2018/07/09/business/starbucks-plastic-straws.html.

10   "Astroturf Organizing 101: Walton Family Gives $6.3 Million to Parent Revolution." *Modern School,* 10 April 2013, http://modeducation.blogspot.com/2013/04/astroturf-organizing-101-walton-family.html.

11   Monbiot, George. *Heat.* Penguin Books, 2006.

Chapter 3: More Than Just Green

1   Celia Santos, Arnaldo Coelho, Alzira Marques, "Does Greenwashing Affect Employee's Career Satisfaction? The Mediating Role of Organizational Pride, Negative Emotions and Affective Commitment?" *Research Square,* 2022, https://doi.org/10.21203/rs.3.rs-1197221/v2.

PART 2: THE CORPORATE SECTOR

Chapter 4: The Good, the Bad, and the Ugly

1   David Carrig, "The US Used To Ship 4,000 Recyclable Containers A Day To China. Where Will The Banned Trash Go Now?" *USA Today,* 21 June 2018, https://www.usatoday.com/story/news/nation-now/2018/06/21/china-ban-plastic-waste-recycling/721879002/.

2   Minter, Adam, "Junkyard Planet: Travels in the Billion-Dollar Trash Trade" Bloomsbury Press, 2013.

3   Kenneth P. Pucker, "Overselling Sustainability Reporting." *Harvard Business Review,* May/June 2021, https://hbr.org/2021/05/overselling-sustainability-reporting.

4   Robert G. Eccles and Svetlana Klimenko, "The Investor Revolution." *Harvard Business Review,* May/June 2019, https://hbr.org/2019/05/the-investor-revolution.

5   "2020 Global Marketing Trends." *Deloitte Insights,* 2019, https://www2.deloitte.com/content/dam/insights/us/articles/2020-global-marketing-trends.

Chapter 5: The Un-Sustainable

1   Pearce, Joshua M. "Towards Quantifiable Metrics Warranting Industry-Wide Corporate Death Penalties." *Social Sciences, vol. 8,* no. 2, 2019, p. 62. MDPI, http://doi.org/10.3390/socsci8020062.

2   "Human Rights Impact Assessment." *Imperial Brands,* 2016, https://www.imperialbrandsplc.com/content/dam/imperial-brands/corporate/sustainability/sustainability-documents/human-rights-assesment.pdf.

3 "Journey to a Better Tomorrow." *British American Tobacco*, 2020, https://www.bat.com/group/sites/UK__9D9KCY, .nsf/vwPagesWebLive/DO964UGU/$file/BAT_Human_Rights_Report_2020. pdf.

4 "Reynolds American Deceives Consumers by Marketing American Spirit Cigarettes As 'Eco Friendly.'" *Cision PR NewswIre*, 25 June 2011, https://www.prnewswire.com/news-releases/reynolds-american-deceives-consumers-by-marketing-american-spirit-cigarettes-as-eco-friendly-126117098.html.

5 Ibid.

6 "PwC's Global Aerospace and Defense: Annual Industry Performance and Outlook." *PriceWaterhouseCoopers*, 2021, https://www.pwc.com/us/en/industries/industrial-products/library/aerospace-defense-review-and-forecast.html.

7 "Revenue of Defense Technology Supplier Lockheed Martin by Product Segment from 2010 to 2021." *Statista*, accessed 10 March 2022, https://www.statista.com/statistics/268931/revenue-of-defense-supplier-lockheed-martin-by-product-segment/.

8 Ibid., PwC.

9 Diana Dimitrova, Mike Lyons, Pelayo Losada, Mike Mester, Tina Zuzek, Marine Baudin-Sarlet, and Matthieu Schmitt, "The Growing Climate Stakes for the Defense Industry." *Boston Consulting Group*, 10 September 2021, https://www.bcg.com/publications/2021/growing-climate-stakes-for-the-defense-industry.

10 Ibid.

11 "Human Rights Report." *Lockheed Martin*, 14 October 2021, https://www.lockheedmartin.com/content/dam/lockheed-martin/eo/documents/sustainability/Lockheed_Martin_Human_Rights_Report_2020.pdf.

12 "Naval Group's Vision of Commitment." *Naval Group*, accessed 10 March 2022, https://www.naval-group.com/en/commitment.

13 "Corporate and Social Responsibility Report 2020." *MBDA Missile Systems*, 2020, https://www.mbda-systems.com/wp-content/uploads/2021/10/MBDA_CSR_REPORT_2020.pdf.

14 The Hon Justice Steven Rares, "Ships That Changed the Law - The Torrey Canyon Disaster." *The Federal Court of Australia*, 5 October 2017, https://www.fedcourt.gov.au/digital-law-library/judges-speeches/justice-rares/rares-j-20171005.

15 Ibid.

16 "Oil Tanker Spill Statistics," *The International Tanker Owners Pollution Federation*, accessed 20 March 2022, https://www.itopf.org/knowledge-resources/data-statistics/statistics/.

17 Elliot Negin, "Why is ExxonMobil Still Funding Climate Science Denier Groups?" *The Equation*, 31 August 2018, https://blog.uc-susa.org/elliott-negin/exxonmobil-still-funding-climate-science-denier- groups.

18 Hunt, George. *Toward Self-Sufficiency: A Community for a Transition Period*. Indiana, iUniverse, 2018.

19 Ibid., Negin.

20 David Adam, "Exxon to Cut Funding to Climate Change Denial Groups." *The Guardian*, 28 May 2008, https://www.theguardian.com/environ- ment/2008/may/28/climatechange.fossilfuels.

21 I've done my best to summarize what can only be described as one of the strangest sagas in legal history. The entire thing could fill books and books, so you'll forgive the brevity as I try to describe a 30+ year story in a few pages. My main sources of information included: Erin Brockovich, "This Lawyer Should Be World-Famous For His Battle With Chevron – But He's In Jail." *The Guardian*, 9 February 2022, https://www.theguardian.com/commentisfree/2022/feb/08/chevron-amazon-ecuador-steven-donziger-erin-brockovich; "Steven Donziger." Wikipedia, Wikimedia Foundation, 1 May 2022, https://en.wikipedia.org/wiki/Steven Donziger; and, "Steven Donziger vs. Big Oil." The Intercept, 27 April 2022, https://theintercept.com/2022/04/27/deconstructed-steven-donziger-chevron-ecuador/.

22 Ibid., Brockovich.

23 Donziger, S. [@SDonzier]. "Day 950 of detention. Includes 45 nights in prison…" Twitter, 14 March 2022, 5:07 a.m., https://twitter.com/SDonziger/status/1503070313537581057.

24 "Chevron Announces Fourth Quarter 2021 Results." *Chevron*, 28 January 2022, https://www.chevron.com/newsroom/2022/Q1/chevron-announces-fourth-quarter-2021-results.

25 Damian Carrington and Matthew Taylor, "Revealed: The 'Carbon Bombs' Set To Trigger Catastrophic Climate Breakdown." *The Guardian*, 11 May 2022, https://www.theguardian.com/environment/ng-interactive/2022/may/11/fossil-fuel-carbon-bombs-climate-breakdown-oil-gas?CMP=Share_iOSApp_Other.

26 "Sustainability." *BP*, accessed 11 March 2022, https://www.bp.com/en/global/corporate/sustainability.html.

27 "Advancing a Lower Carbon Future." *Chevron*, accessed 10 March 2022, https://www.chevron.com/sustainability.

28 "Ecuador Lawsuit." *Chevron*, accessed 10 March 2022, https://www.chevron.com/ecuador.

Chapter 6: Those You Can Trust

1   "Understanding the Trust Equation." *TrustedAdvisor*, accessed 12 January 2022, https://trustedadvisor.com/why-trust-matters/understanding-trust/understanding-the-trust-equation.

2   Go ahead…you know you want to! https://trustsuite.trustedadvisor.com/.

3   Ben Baldanza, "The 1,500-Hour Rule Has Broken The Pilot Pipeline In The U.S." *Forbes*, 11 July 2022, https://

www.forbes.com/sites/
benbaldanza/2022/07/11/the-
1500-hour-rule-has-broken-
the-pilot-pipeline-in-the-us/.

4   Jasmyne Jeffery, "What Are The
Chances Of A Plane Crashing
In 2022?" *HITC*, November
2022, https://www.hitc.com/en-
gb/2022/10/26/what-are-the-
chances-of-a-plane-crashing-
in-2022/.

5   Unpublished confidential
document, *Hotspex*, 28
September 2022.

6   For the purposes of this
very unscientific exercise,
I've included the following
industries: agriculture,
automotive, aviation,
construction, consumer-
packaged goods (food),
consumer-packaged goods
(household), defense, fast
fashion, fast food, fast-moving
consumer goods, finance, gold
and mining, grocery, health
and beauty, medical technology,
oil and gas, paper goods,
pharmaceuticals, shipping,
technology, tobacco, and travel
and tourism.

7   Emma Shepherd, "Woolies and
Bunnings Remain Australia's
Most Trusted Brands."
Mumbrella, 24 October 2022,
https://mumbrella.com.au/
woolies-and-bunnings-remain-
australias-most-trusted-
brands-761736.

8   "Rankings: Global RepTrak
100." *RepTrack*, accessed 12
January 2022, https://www.
reptrak.com/rankings/.

9   "Nike Sweatshops: Inside The
Scandal." *New Idea*, accessed
20 February 2023, https://
www.newidea.com.au/nike-
sweatshops-the-truth-about-
the-nike-factory-scandal.

Chapter 7: Those That Can Change

1   The Hon Justice Steven Rares,
"Ships that Changed the Law -
The Torrey Canyon Disaster."
Federal Court of Australia, 5
October 2017, https://www.
fedcourt.gov.au/digital-law-
library/judges-speeches/justice-
rares/rares-j-20171005.

2   "Why the 'S' in ESG is Still in
Focus." *Merrill*, 16 September
2022, https://www.ml.com/
articles/why-social-is-
important-to-esg-investing.
html.

3   "How a Purpose-Driven
Workplace Impacts the
Employee Experience." *Rise*, 12
January 2021, https://risepeople.
com/blog/purpose-driven-
workplace/.

4   "A 'Natural' Rise in
Sustainability Around the
World." *Nielsen IQ*, 10 January
2019, https://nielseniq.
com/global/en/insights/
analysis/2019/a-natural-rise-
in-sustainability-around-the-
world/.

5   "APAC Post-COVID-19
Consumer Trends Will Bring
Re-Calibration Of Sustainability
Priorities, Says GlobalData."
GlobalData, 4 June 2020,
https://www.globaldata.com/
media/consumer/apac-post-
covid-19-consumer-trends-
will-bring-re-calibration-of-
sustainability-priorities-says-
globaldata/.

6   Fernando Belinchón and
Qayyah Moynihan, "25 Giant

Companies That Are Bigger Than Entire Countries." *Business Insider España*, 25 July 2018, https://www.businessinsider.com/25-giant-companies-that-earn-more-than-entire-countries-2018-7.

7   Docusign. "Paper is Costing Us the Earth: A Guide of Sustainable Measures for Better Business." November 2022, https://www.docusign.com.au/white-papers/is-paper-costing-us-the-earth?utm_campaign=APJ_AN_CMM_AWA_2210_PaperCostingEarth.

8   "Environmental, Social, and Governance." *DocuSign*, accessed 11 January 2023, https://investor.docusign.com/investors/ESG/.

9   "Electromobility." *BMW Group*, accessed 11 January 2023, https://www.bmwgroup.com/en/sustainability.html.

10   Kripa Jayaram, Chris Kay, and Dan Murtaugh, "China Reduced Air Pollution in 7 Years as Much as US Did in Three Decades." *Bloomberg*, 14 June 2022, https://www.bloomberg.com/news/articles/2022-06-14/china-s-clean-air-campaign-is-bringing-down-global-pollution.

11. 11   For more on Walmart's Women in Factories Program, including ongoing progress updates, see: https://www.bsr.org/en/collaboration/groups/women-in-factories-china-program.

12   Christine Lellis, "17 Major Companies Linking Executive Pay to ESG Performance." *Perillon*, 6 April 2021, https://www.perillon.com/blog/17-major-companies-linking-executive-pay-to-esg-performance.

13   "ESG + Incentives 2021 Report." *Semler Brossy*, 2021, https://semlerbrossy.com/wp-content/uploads/2021/06/SemlerBrossy-ESG-Report-Issue-1-2021.pdf.

14   Maria Castañón Moats, Leah Malone, and Christopher Hamilton, "The Evolving Role of ESG Metrics in Executive Compensation Plans." *Harvard Law School Forum on Corporate Governance*, 19 March 2022, https://corpgov.law.harvard.edu/.

15   Ibid.

16   "Most Diverse Boards." *Bank Director*, accessed 10 January 2023, https://www.bankdirector.com/rankingbanking/most-diverse-boards/.

17   Shannon Cabral and Daniel Krasner, "The 5 JUST 100 Companies Leading on Gender Board Diversity." *Just Capital*, 23 May 2022, https://justcapital.com/news/top-companies-for-board-gender-diversity-2022/.

18   "Gender Diversity Index of Women on Boards and in Corporate Leadership." *European Women on Boards*, 2021, https://europeanwomenonboards.eu/wp-content/uploads/2022/01/2021-Gender-Diversity-Index.pdf.

19   Molly Stutzman, "A Growing Number of Companies Are Recognizing the Benefits of Racially Diverse Boards." *Just Capital*, accessed 10 January 2023, https://justcapital.com/

reports/a-growing-number-of-companies-and-shareholders-are-recognizing-the-benefits-of-racially-diverse-boards/.

20 Julie Sweet, "Continuing Our Progress – Taking Action Against Racism." LinkedIn, 2 October 2020, https://www.linkedin.com/pulse/continuing-our-progress-taking-action-against-racism-julie-sweet/.

21 "Board Diversity Statistics." *The Australian Institute of Company Directors,* accessed 9 January 2023, https://www.aicd.com.au/about-aicd/governance-and-policy-leadership/board-diversity/Board-diversity-statistics.html

22 Hannah Wootton, "Nearly 90pc of IPO Boards Are All-Male: New Data." *Australian Financial Review,* 23 September 2022, https://www.afr.com/politics/federal/nearly-90pc-of-ipo-boards-are-all-male-new-data-20220916-p5biq5.

23 Ibid., European Women on Boards.

24 "Board Diversity in 2021." Mogul, 2022, https://hub.onmogul.com/hubfs/Board-Diversity-in-2022.pdf.

25 Ibid

26 "Diversity." *TrueUp Tech,* accessed 11 January 2023, https://www.trueup.io/companies/diversity.

27 Federica Urso, "Number of Company Sustainability Officers Triples in 2021 – Study." *Reuters,* 5 May 2022, https://www.reuters.com/business/sustainable-business/number-company-sustainability-officers-triples-2021-study-2022-05-04/.

28 "Top 100 Sustainability Leaders 2022." *Sustainability Magazine,* accessed 11 January 2023, https://sustainabilitymag.com/magazine/top-100-sustainability-leaders.

Chapter 8: Fast Fashion

1 Ella Moffat and Smita Nimilita, "What If All Garment Workers in Bangladesh Were Financially Included?" *BSR,* 19 January 2022, https://www.bsr.org/en/our-insights/blog-view/what-if-all-garment-workers-in-bangladesh-were-financially-included.

2 Ruma Paul and Krishna Das, "Bangladesh Garment Export Growth Seen Slowing to 'Normal' 15% This Year." *Reuters,* 10 August 2022, https://www.reuters.com/markets/asia/bangladesh-garment-export-growth-seen-slowing-normal-15-this-year-2022-08-10/.

3 Tim McDonnell, "Climate Change Creates a New Migration Crisis for Bangladesh." *National Geographic,* 24 January 2019, https://www.nationalgeo-graphic.com/environment/2019/01/climate-change-drives-migration-crisis-in-bangladesh-from-dhaka-sundabans/.

4 Jason Burke, "Rana Plaza: One Year On From The Bangladesh Factory Disaster." *The Guardian,* 19 April 2014, https://www.theguardian.com/world/2014/apr/19/rana-plaza-bangladesh-one-year-on.

5   Dana Thomas, "The High Price of Fast Fashion." *The Wall Street Journal*, 29 August 2019, https://www.wsj.com/articles/the-high-price-of-fast-fashion-11567096637.

6   "Survey: Women's Closets Are Full to the Brim." *ClosetMaid*, accessed March 9, 2022, https://blog.closetmaid.com/2016/05/full-to-the-brim.

7   It was a stinky job, too. For more on the entire process of creating Tyrian Purple, read David Jacoby's 2004 account in: "Silk Economics and Cross-Cultural Artistic Interaction: Byzantium, the Muslim World, and the Christian West." *Dumbarton Oaks Papers*, 2004, pp. 197–240. doi:10.2307/3591386.

8   The history of the sewing machine is laced with controversy. A timeline of the drama can be found in: "History of the Sewing Machine: A Story Stitched in Scandal." *Contrado*, 6 March 2019, https://au.contrado.com/blog/history-of-the-sewing-machine/.

9   Meryl Baer, "The History of American Income." *Bizfluent*, 26 September 2017, https://bizfluent.com/info-7769323-history-american-income.html.

10   Lynn Yaeger, "Remembering the Lessons of the Triangle Shirtwaist Factory Fire, 106 Years Later." *Vogue*, 25 March 2017, https://www.vogue.com/article/triangle-shirtwaist-factory-106-anniversary.

11   "On This Day: March 25, 1911: The Triangle Shirtwaist Fire." *The Sheila Variations*, accessed 11 March 2022, http://www.sheilaomalley.com/?p=35627.

12   Brian Knutson, Scott Rick, G. Elliott Wimmer, Drazen Prelec, and George Loewenstein. "Neural Predictors of Purchases." *Neuron*. 2007, pp. 147-156. https://www.ncbi.nlm.nih.gov/pmc/articles/PMC1876732/.

13   "UN Helps Fashion Industry Shift to Low Carbon." *United Nations Climate Change*, 6 September 2018, https://unfccc.int/news/un-helps-fashion-industry-shift-to-low-carbon.

14   Christine Ro, "Can Fashion Ever Be Sustainable?" *BBC Future*, 11 March 2020, https://www.bbc.com/future/article/20200310-sustainable-fashion-how-to-buy-clothes-good-for-the-climate.

15   Ibid., United Nations Climate Change.

16   Food and Agriculture Organization of the United Nations, "Emissions Due to Agriculture." FAOSTAT Analytical Brief Series, 2020, https://www.fao.org/3/cb3808en/cb3808en.pdf.

17   Ibid., United Nations Climate Change.

18   "Fast Fashion's Massive Waste Problem." *City of Greater Geelong*, accessed 23 May 2022, https://www.geelongaustralia.com.au/recycling/article/item/8d48e38cacff8ae.aspx.

19   Rachel Brown, "The Environmental Crisis Caused By Textile Waste." *Roadrunner*, 8 January 2021, https://www.roadrunnerwm.com/blog/textile-waste-environmental-crisis.

20 Abigail Beall, "Why Clothes Are So Hard to Recycle." *BBC Future*, 13 July 2020, https://www.bbc.com/future/article/20200710-why-clothes-are-so-hard-to-recycle.

21 "Let's Close the Loop." *H&M*, accessed 25 May 2022, https://www2.hm.com/en_au/sustainability-at-hm/our-work/close-the-loop.html.

22 Ibid., Beall.

23 Ibid., Brown.

24 Ibid., Beall.

25 If you're interested in jumping on the consignment bandwagon, there are plenty of websites to check out. Some of the most popular are The RealReal, Depop, and ThredUP. Many global brands also have their own resell sites, but make sure to do your research before using some of them. The most reputable are from Patagonia, Eileen Fisher, and REI.

26 "Circular Business Models: Redefining Growth for a Thriving Fashion Industry." *The Ellen MacArthur Foundation*, accessed 25 May 2022, https://ellenmacarthurfoundation.org/fashion-business-models/overview.

27 "Our Quest for Circularity." *Patagonia*. accessed 25 May 2022, https://www.patagonia.com.au/blogs/roaring-journals/our-quest-for-circularity.

28 Kristen Faranakis, "Fast Fashion's 'Sustainability' Endeavors Need to be About More Than Fabrics, Recycling." *The Fashion Law*, 2 April 2021, https://www.thefashionlaw.com/fast-fashion-sustainability-is-about-more-than-the-fabrics/.

29 I wonder if this piece is still shown in business schools today? Ten years on, it's definitely not stood the test of time. Sull, Donald & Turconi, Stefano. "Fast Fashion Lessons." *Business Strategy Review, vol. 19*, 2008, pp. 4-11. 10.1111/j.1467-8616.2008.00527.x.

30 Channing Hargrove, "Zara Announces Sustainability Initiatives — But What About Its Factory Workers?" *Refinery29*, 17 July 2019, https://www.refin-ery29.com/en-gb/zara-sustainable-initiatives.

31 Ibid., Burke.

32 Sarah Butler, "Bangladesh Clothing Factory Safety Deal In Danger, Warn Unions." *The Guardian*, 22 April 2021, https://www.theguardian.com/world/2021/apr/22/bangladesh-clothing-factory-safety-deal-in-danger-warn-unions.

PART 3: STATE-SPONSORED GREENWASHING

Chapter 11: COP Out

1 "Summary report, 3–14 November 2008." *IISD Earth Negotiations Bulletin,* accessed 15 January 2022, https://enb.iisd.org/events/unccd-cric-7-and-1st-special-session-committee-science-and-technology-cst-s-1/summary-report#analysis-heading.

2 If all these acronyms weren't bad enough, the UN loves oscillating between British and American English, with the occasional French word thrown in for fun.

3   Alister Doyle, "Nations Call For Tougher U.N. Environment Role." *Reuters*, 4 February 2007, https://www.reuters.com/article/uk-globalwarming-appeal-idUKL0335755320070203.

4   Glenn Beck and Harriet Parke, "Agenda 21." Threshold Editions (2013) accessed 12 January 2022, https://www.google.com.au/books/edition/Agenda_21/ZXwVAAAAQBAJ?hl=en&gbpv=0. For those wondering, I pulled the quote from a Google Book intro because it'll be a cold day in Hell before I give this man any of my money.

5   "Canada pulls out of Kyoto Protocol." *CBC News*, 12 December 2011, https://www.cbc.ca/news/politics/canada-pulls-out-of-kyoto-protocol-1.999072.

6   The Agreement itself is pretty short, and definitely worth a read to learn more. You can find it at: https://unfccc.int/sites/default/files/english_paris_agreement.pdf.

7   Robinson Meyer, "The Most Stupendous Acronym From the Paris Climate Talks." *The Atlantic*, 10 December 2015, https://www.theatlantic.com/science/archive/2015/12/the-most-stupendous-acronym-from-the-paris-climate-talks/419631/.

8   "History of UN Climate Talks." *Center for Climate and Energy Solutions*, accessed 17 January 2023, https://www.c2es.org/content/history-of-un-climate-talks.

9   Damian Carrington, "Climate Tipping Points Could Topple Like Dominoes, Warn Scientists." *The Guardian*, 4 June 2021, https://www.theguardian.com/environment/2021/jun/03/climate-tipping-points-could-topple-like-dominoes-warn-scientists.

10 Ruth Michaelson, "'Explosion' In Number Of Fossil Fuel Lobbyists At Cop27 Climate Summit." *The Guardian*, 10 November 2022, https://www.theguardian.com/environment/2022/nov/10/big-rise-in-number-of-fossil-fuel-lobbyists-at-cop27-climate-summit.

11 Nathan Cooper and Amy White, "IPCC Report: Urgent Climate Action Needed To Halve Emissions By 2030." *World Economic Forum*, 6 April 2022, https://www.weforum.org/agenda/2022/04/ipcc-report-mitigation-climate-change/.

12 "Life Under the Ozone Hole." *Newsweek*, 8 December 1991, https://www.newsweek.com/life-under-ozone-hole-201054.

13 "United Nations Charter (Full Text)." *United Nations*, accessed 14 January 2022, https://www.un.org/en/about-us/un-charter/full-text.

Chapter 12: Toeing the Line

1   "What is THE LINE?" Youtube, uploaded by NEOM, 26 July 2022, https://www.youtube.com/watch?v=0kz5vEqdaSc.

2   Anuz Thapa, "Saudi Arabia's $500 billion bet to build a futuristic city in the desert." *CNBC*, 14 Jan 2023, https://

www.cnbc.com/2023/01/13/neom-is-saudi-arabias-500-billion-bet-to-build-a-futuristic-city-.html.

3   Ruth Michaelson, "'It's being built on our blood': the true cost of Saudi Arabia's $500bn megacity." *The Guardian*, 4 May 2020, https://www.theguardian.com/global-development/2020/may/04/its-being-built-on-our-blood-the-true-cost-of-saudi-arabia-5bn-mega-city-neom.

4   "Death Sentences For Men Who Refused To Make Way For Neom." *ALQST*, 10 October 2022, https://www.alqst.org/en/post/death-sentences-for-men-who-refused-to-make-way-for-neom.

5   Tom Ravenscroft, "Architecture Studios "Benefiting" From Alleged Human Rights Violations At Neom." *DeZeen*, 13 December 2022, https://www.dezeen.com/2022/12/13/human-rights-violations-neom-amnesty-international/.

6   "Saudi Arabia Oil." *Worldometer,* accessed 19 January 2023, https://www.worldometers.info/oil/saudi-arabia-oil/.

7   I was shocked to learn the "country" with the largest per-capita use of oil was Gibraltar at 34,000 gallons. Compare this to number two, Singapore, at 3,679 gallons. I'm not sure why Gibraltar uses ten times more oil than Singapore and have no interest in opening that can of worms.

8   Nishant Ugal, "No Capacity: Saudi Arabia Cannot Expand Oil Production Beyond 13 Million Bpd." *Upstream,* 20 July 2022, https://www.upstreamonline.com/production/no-capacity-saudi-arabia-cannot-expand-oil-production-beyond-13-million-bpd/2-1-1262005.

9   "Countries (Updated December 2022)." *Climate Action Tracker,* accessed 5 January 2023, https://climateactiontracker.org/countries/.

10 Matthew Hall, "Adani's Carmichael Coal Mine Controversy Explained." *Mining Technology,* 20 July 2020, https://www.mining-technology.com/features/adani-carmichael-controversy-explained/.

11 Ibid.

12 "Australian Mineral Facts." *Geoscience Australia,* accessed 20 January 2023, https://www.ga.gov.au/education/classroom-resources/minerals-energy/australian-mineral-facts.

13 "Australia - Country Commercial Guide." *International Trade Administration,* accessed 20 January 2023, https://www.trade.gov/country-commercial-guides/australia-mining.

14 Nnamdi Anyadike, "Lessons Not Learned: Western Australia In Clash With Indigenous Peoples' Rights." *Mining Technology,* 6 July 2022, https://www.mining-technology.com/features/western-australia-mining-indigenous/.

15 Ibid, Hall.

16 Polly Hemming, "How the Government Supports Greenwashing." *The Saturday Paper,* 29 October 2022, https://www.thesaturdaypaper.com.au/

environment/life/2022/10/29/
how-the-government-supports-
greenwashing#hrd.

17   Ibid.

18   For a deeper dive into just how
bad things have gotten, check
out Australia's State of the
Environment Report, released
in 2022. It's a laundry list of
failures from a willfully ignorant
Government.

19   "Corruption Perceptions Index:
Singapore." *Transparency
International*, accessed 20
January 2023, https://www.
transparency.org/en/cpi/2021/
index/sgp.

20   "Life Expectancy of the World
Population." *Worldometer*,
accessed 20 January 2023,
https://www.worldometers.info/
demographics/life-expectancy/.

21   "Mortality rate, under-5
(per 1,000 live births) –
Singapore." *The World Bank*,
accessed 20 January 2023,
https://data.worldbank.
org/indicator/SH.DYN.
MORT?locations=SG&most_
recent_value_desc=false.

22   Nigel Chua, "Mothership
Explains: Why Most S'poreans
Own Their Homes Instead
Of Renting." *Mothership*,
28 September 2021, https://
mothership.sg/2021/09/hdb-
home-ownership-history/.

23   Garry Lu, "Which Countries
Have The World's Fastest
Internet Speeds?" *Boss Hunting*,
27 September 2021, https://
www.bosshunting.com.au/
lifestyle/technology/worlds-
fastest-internet-speeds/.

24   Justin Ong, "S'pore Population
Better Educated Across Age,

Ethnicity; Women Make
Greater Strides." *Straits Times*,
16 June 2021, https://www.
straitstimes.com/singapore/
spore-population-better-
educated-across-age-ethnicity-
women-make-greater strides.

25.   "Singapore a City in a Garden
– A Model For Creating An
Integrated Urban Green
Walking Network." *Natural
Walking Cities*, accessed
20 January 2023, https://
naturalwalkingcities.com/
singapore-a-city-in-a-garden-
a-model-for-creating-an-
integrated-urban-green-
walking-network/.

26.   "A City of Green Possibilities."
SG Green Plan, accessed 20
January 2023, https://www.
greenplan.gov.sg/.

27   "Country Summary: Singapore."
Climate Action Tracker, accessed
20 January 2023, https://
climateactiontracker.org/
countries/singapore/.

28.   "Populist Parties in Europe Are
Now Talking About Climate
Change." *Marsh McLennen
BRINK*, 1 February 2022,
accessed 20 January 2023,
https://www.brinknews.com/
populist-parties-in-europe-are-
now-talking-about-climate-
change/.

29.   Jake Horton and Daniele
Palumbo, "Russia sanctions:
What impact have they had on
its oil and gas exports?" *BBC
News*, 24 January 2023, https://
www.bbc.com/news/58888451.

30.   Lauri Myllyvirta, "EU Ban On
Russian Oil: Why It Matters
And What's Next." *CREA*,
5 December 2022, https://

energyandcleanair.org/eu-ban-on-russian-oil-why-it-matters-and-whats-next.

31. "Germany Says It Is No Longer Reliant On Russian Energy." *BBC News,* 20 January 2023, https://www.bbc.com/news/business-64312400.

32. Simon Kuper, "The Netherlands May Be The First Country To Hit The Limits Of Growth." *Financial Times,* 27 October 2022, https://www.ft.com/content/4c56c9b2-f4ad-4956-9216-655acebd845d.

33. Erik Hoffner, "How Unsustainable Is Sweden's Forestry? 'Very.' Q&A With Marcus Westberg And Staffan Widstrand." *Mongabay,* 17 June 2022, https://news.mongabay.com/2022/06/how-unsustainable-is-swedens-forestry-very-qa-with-marcus-westberg-and-staffan-widstrand/.

Chapter 13: Hyper-National Organisations

1   Pete Pattisson, Niamh McIntyre, Imran Mukhtar, Nikhil Eapen, Imran Mukhtar, Md Owasim Uddin Bhuyan, Udwab Bhattarai, and Aanya Piyari, "Revealed: 6,500 Migrant Workers Have Died In Qatar Since World Cup Awarded." *The Guardian,* 23 February 2021, https://www.theguardian.com/global-development/2021/feb/23/revealed-migrant-worker-deaths-qatar-fifa-world-cup-2022.

2   Ian Ward, "The Many, Many Controversies Surrounding The 2022 World Cup, Explained." *Vox,* 19 November 2022, https://www.vox.com/world/23450515/world-cup-fifa-qatar-2022-controversy-scandals-explained.

3   Jules Boykoff, "The World Cup In Qatar Is a Climate Catastrophe." *Scientific American,* 23 November 2022, https://www.scientificamerican.com/article/the-world-cup-in-qatar-is-a-climate-catastrophe/.

4   Ibid., Ward.

5   Müller, M., Wolfe, S.D., Gaffney, C. et al. "An Evaluation Of The Sustainability Of The Olympic Games." *Nat Sustain,* 4, 340–348 (2021), https://doi.org/10.1038/s41893-021-00696-5.

6   "About Sport in Australia." *Department of Health and Aged Care,* accessed 23 January 2023, https://www.health.gov.au/topics/sport/about-sport-in-australia.

7   "Out Of Bounds: Coal, Gas And Oil Sponsorship In Australian Sports." *Australian Conservation Foundation,* October 2022, https://assets.nationbuilder.com/auscon/pages/21088/attachments/original/1666744415/Out_of_Bounds_Report_October_2022.pdf?1666744415.

8   Ibid.

9   Ibid.

10   "Sweat Not Oil: Why Sports Should Drop Advertising and Sponsorship From High-Carbon Polluters." *New Weather Institute,* March 2021, https://static1.squarespace.com/static/5ebd0080238e863d04911b51/t/605b60b09a957c1b05f433e2/1616601271774/

Sweat+Not+Oil+-+why+Sport
s+should+drop+advertising+fr
om+high+carbon+polluters+-
+March+2021v3.pdf.

11　Dan Falkenheim and Alex
Prewittnov, "What on Earth:
How Phony Environmentalism
Came to Sports." *Sports
Illustrated*, 1 November
2022, https://www.si.com/
soccer/2022/11/01/sports-
greenwashing-daily-cover.

12　Ibid.

13　Ibid., *Australian Conservation
Foundation*.

14　Ibid., *New Weather Institute*.

15　Ibid., *Australian Conservation
Foundation*.

16　Seth Wynes. "COVID-19
Disruption Demonstrates Win-
Win Climate Solutions for
Major League Sports."
*Environmental Science &
Technology*, 2021 55 (23),
15609-15615, DOI: 10.1021/acs.
est.1c03422.

17　Australia Institute, "Majority
Agree with Banning Fossil
Fuel Sponsorship in Sport:
Research Polling." *Mirage News*,
24 October 2022, https://www.
miragenews.com/majority-
agree-with-banning-fossil-
fuel-880632/.

18　Tony Sokol, "Veep Season 7
Episode 2 Review: Discovery
Weekend." *Den of Geek*, 8 April
2019, https://www.denofgeek.
com/tv/veep-season-7-episode-
2-review-discovery-weekend/.

19　Amanda Shendruk, "Here are
the countries and companies
dominating Davos." *Quartz*, 15
January 2023, https://qz.com/
davos-2023-world-economic-
forum-attendees-1849990706;

and, Siddharth K., "Davos
2023: The World Economic
Forum Explained." *Reuters*,
17 January 2023, https://www.
reuters.com/business/davos-
2023-world-economic-forum-
explained-2023-01-16/.

20　"World Economic Forum
Annual Meeting." *World
Economic Forum*, accessed 25
January 2023, https://www.
weforum.org/events/world-
economic-forum-annual-
meeting-2023/.

21　Ibid.

22　"Al Gore at DAVOS 2023: "We
Have the Tech to Slow Climate
Change – So Where's Political
Will." *World Economic Forum*,
accessed 25 January 2023,
https://www.weforum.org/
videos/we-have-the-tech-to-
slow-climate-change-so-where-
s-political-will-asks-al-gore.

23　Kaya Williams, "At Aspen
Ideas Festival, A Search For
'Environmental Identity'."
The Aspen Times, 2 July 2022,
https://www.aspentimes.com/
news/at-aspen-ideas-festival-
a-search-for-environmental-
identity/.

24　"Davos Attendees Can't Stop
Hating on Gen Z." *Bloomberg*,
21 January 2023, https://
www.bloomberg.com/news/
videos/2023-01-20/davos-
attendees-can-t-stop-hating-on-
gen-z.

25　Maddie Bender, "Exxon's Own
Science Was Scary Accurate
About Global Warming. So It
Covered It Up." *Yahoo! Finance*,
13 January 2023, https://
finance.yahoo.com/news/
exxon-own-science-scary-
accurate-094851864.html.

26 Cloe Read, "Adani and Prestigious London Museum Cop Heat After Teaming Up On Climate Project." *Brisbane Times*, 20 October 2021, https://www.brisbanetimes.com.au/national/queensland/adani-and-prestigious-london-museum-cop-heat-after-teaming-up-on-climate-project-20211020-p591li.html.

27 Matthew Taylor, "Dozens Of Academics Shun Science Museum Over Fossil Fuel Ties." *The Guardian,* 20 November 2021, https://www.theguardian.com/culture/2021/nov/19/dozens-of-academics-shun-science-museum-over-fossil-fuel-ties.

28 Matthew Taylor, "Hundreds of Teachers Boycott Science Museum Show Over Adani Sponsorship." *The Guardian,* 15 July 2022, https://www.theguardian.com/culture/2022/jul/15/hundreds-of-teachers-boycott-science-museum-over-adani-sponsorship.

29 Ibid.

30 "Palm Oil Scorecard." *WWF,* accessed 24 January 2023, http://palmoilscorecard.panda.org/#/home.

31 Russel Hargrave, "Charities Could Be Used By Corporate Partners For 'Greenwashing', Report Warns." *Third Sector,* 20 September 2022, https://www.thirdsector.co.uk/charities-used-corporate-partners-greenwashing-report-warns/fundraising/article/1799383.

32 "Kulim (Malaysia) Berhad Supports Land Trust's Orangutan Conservation Initiative." *Kabar Medan,* 29 September 2015, https://kabarmedan.com/kulim-malaysia-berhad-dukung-inisiatif-konservasi-orangutan-land-trust/.

33 "Greenwashing Tactic #9: Partnerships, Sponsorships & Research Funding." *Palm Oil Detectives,* 14 October 2021, https://palmoildetectives.com/2021/10/14/greenwashing-tactic-9-partnerships/#OLT-accepts-cheque-for-500K.

34 Matthew Taylor, "Royal Shakespeare Company to End BP Sponsorship Deal." *The Guardian,* 3 October 2019, https://www.theguardian.com/stage/2019/oct/02/royal-shakespeare-company-to-end-bp-sponsorship-deal.

PART 4: INFLUENCES

Chapter 16: Activists, Moralists, and NGOs...Oh My!

1 Dejan Jotanovic, "Why do most people find us vegans so annoying?" *The Guardian,* 10 September 2019, https://www.theguardian.com/commentisfree/2019/sep/10/why-do-most-people-find-us-vegans-so-annoying.

2 "Angel." *Genius,* accessed 1 February 2023, https://genius.com/Sarah-mclachlan-angel-lyrics.

3 Samantha Lefave, "Sarah McLachlan Reveals the Truth About Those Sad ASPCA Ads." *Redbook,* 5 January 2016, https://www.redbookmag.com/life/pets/news/a41805/sarah-mclachlan-aspca-commercial/.

4   Ibid.

5   Sophie Russell, "Opinion: Why 'Sadvertising' And Melancholy Marketing Doesn't Help Charities." CharityTimes, 8 May 2019, https://www.charitytimes.com/ct/blog-why-sadvertising-doesnt-help-charities.php.

6   South Park. 2015. "Safe Space." South Park Studios. 23 minutes. 21 October 2015. https://www.southparkstudios.com/episodes/oeajfq/south-park-safe-space-season-19-ep-5.

Chapter 17: Bright Lights Cast Long Shadows

1   "Rusty Radiator Award 2015." Radi-Aid, accessed 1 February 2023, https://www.radiaid.com/radi-aid-awards-2017.

2   "Rusty Radiator Award 2015." Radi-Aid, accessed 1 February 2023, https://www.radiaid.com/rusty-radiator-award-2015.

3   Alexandra Schwartz, "The Rose Reading Room and the Real Meaning of 'Luxury' in New York City." The New Yorker, 7 October 2016, https://www.newyorker.com/culture/cultural-comment/nypl-rose-reading-room-and-the-real-meaning-of-luxury-in-new-york-city.

4   Jacob Davidson, "The 10 Richest People of All Time." Money, 30 July 2015, https://money.com/the-10-richest-people-of-all-time-2.

5   "The Rockefellers: A Century of Giving." Rockefeller Philanthropy Advisors, accessed 1 February 2023, https://www.rockpa.org/guide/rockefellers-legacy-giving/.

6   Fang Block, "The Global Ultra Rich Donated $175 Billion in 2020." Barrons, 27 January 2022, https://www.barrons.com/articles/the-global-ultra-rich-donated-175-billion-in-2020-01643322727.

7   "Foundation Fact Sheet." The Bill & Melinda Gates Foundation, accessed 1 February 2023, https://www.gatesfoundation.org/about/foundation-fact-sheet.

8   The dollar amounts, on both the giving and net-worth sides of the equation, are constantly shifting. Figures in this section are from: Máximo Tuja, "America's Top Givers 2022: The 25 Most Philanthropic Billionaires." Forbes, 19 January 2022, https://www.forbes.com/sites/forbeswealthteam/2022/01/19/americas-top-givers-2022-the-25-most-philanthropic-billionaires/.

9   Leon Langdon, "Jeff Bezos: Billionaire, Philanthropist, Greenwasher?" New York University Journal of Political Inquiry, 17 November 2022, https://jpinyu.com/2022/11/17/jeff-bezos-billionaire-philanthropist-greenwasher.

10   Justine Calma, "Bezos' Climate Fund Faces A Reckoning With Amazon's Pollution." The Verge, 5 February 2021, https://www.theverge.com/2021/2/4/22266225/jeff-bezos-climate-change-earth-fund-amazon-pollution.

11   Anmar Frangoul, "Elon Musk Is Smart — But He Doesn't Understand ESG, Tech CEO Says." CNBC, 1 July 2022, https://www.cnbc.

com/2022/07/01/elon-musk-is-a-smart-person-but-he-doesnt-understand-esg-tech-ceo.html.

12 Stephanie Skora, "Pride With Prejudice: Exposing A Wider Bridge's Right-Wing Funding." Accessed 1 February 2022, https://static1.squarespace.com/static/5b572cf60dbda33f837c0ae1/t/5c72f3b19b747a1582c318b4/1551037363459/Pride+With+Prejudice.pdf.

13 Amrith Ramkumar, "Koch Industries, Built on Oil, Bets Big on U.S. Batteries." *The Wall Street Journal,* 22 March 2022, https://www.wsj.com/articles/koch-industries-built-on-oil-bets-big-on-u-s-batteries-11647946147.

14 Ian Simpson, "U.S. Businessman, Philanthropist Jon Huntsman Sr. Dead At 80." *Reuters,* 3 February 2018, https://www.reuters.com/article/us-people-huntsman-idUSKBN1FN00Z.

15 Sam Haddad, "The Problem With Climate Mega-Philanthropy." *Raconteur,* 9 January 2023, https://www.raconteur.net/climate-crisis/the-problem-with-climate-mega-philanthropy/.

16 All products and prices taken from Erewhon's online shop, accessed 1 February 2023, https://shop.erewhonmarket.com/.

17 Daisy Murray, "Oscars 2020: Sustainable Fashion Was The Real Winner Of The 92nd Academy Awards." *Elle,* 10 February 2020, https://www.elle.com/uk/fashion/a30847184/oscars-sustainable-red-carpet/.

18 Hayley C. Cuccinello, "Inside The $9,500 Golden Globes Gift Bag." *Forbes,* 4 January 2019, https://www.forbes.com/sites/hayleycuccinello/2019/01/04/inside-the-nearly-10000-golden-globes-gift-bag.

19 "Flight Carbon Footprint Between Honolulu and Los Angeles," *Curb6,* accessed 2 February 2023, https://curb6.com/footprint/flights/honolulu-hnl/los-angeles-lax.

20 Marc Maslin, "Which Diet Will Help Save Our Planet: Climatarian, Flexitarian, Vegetarian Or Vegan?" *The Conversation,* 12 August 2022, https://theconversation.com/which-diet-will-help-save-our-planet-climatarian-flexitarian-vegetarian-or-vegan.

21 Allyson Chiu, "Celebrities use private jets excessively. It's a climate nightmare." *The Washington Post,* 2 August 2022, https://www.washingtonpost.com/climate-environment/2022/08/02/taylor-swift-kylie-jenner-private-jet-emissions/.

22 Ibid.

23 Yard, "Just Plane Wrong: Celebs with the Worst Private Jet Co2 Emissions." *Yard,* 29 July 2022, https://weareyard.com/insights/worst-celebrity-private-jet-co2-emission-offenders.

24 Ibid.

25 Ibid., Chiu.

26 "Close Up: Carbon Emissions of Film and Television Production." *Sustainable Production Alliance,* March 2021, https://www.greenproductionguide.com/

wp-content/uploads/2021/04/
SPA-Carbon-Emissions-Report.
pdf.

27 California Integrated
Waste Management Board,
"Sustainability in the Motion
Picture Industry." *UCLA
Institute of the Environment*,
November 2006, https://www.
ioes.ucla.edu/wp-content/
uploads/mpisreport.pdf.

28 Abbey White, "Golden
Globes: Read Ricky Gervais'
Scathing Opening Monologue."
The Hollywood Reporter, 5
January 2020, https://www.
hollywoodreporter.com/news/
general-news/transcript-ricky-
gervais-golden-globes-2020-
opening-monologue-1266516/.

29 Olivia Petter, "Fast Fashion:
Boohoo And Missguided
Among Worst Offenders In
Sustainability Inquiry." *The
Independent*, 30 January 2019,
https://www.independent.co.uk/
life-style/fashion/fast-fashion-
boohoo-missguided-brands-
sustainability-environmental-
audit-committee-
2019-a8754496.html.

30 Caroline Wheeler, Amardeep
Bassey, and Vidhathri Matety,
"Boohoo: Fashion Giant Faces
'Slavery' Investigation." *The
Sunday Times*, 5 July 2020,
https://www.thetimes.co.uk/
article/boohoo-fashion-giant-
faces-slavery-investigation-
57s3hxcth.

31 "Annual Report and Accounts,
2022." *Boohoo Group*, 2022,
https://www.boohooplc.
com/sites/boohoo-corp/
files/2022-05/boohoo-com-plc-
annual-report-2022.pdf.

32 "Kourtney Kardashian
Barker's Sustainable Fashion
Journey." *Youtube*, uploaded
by Boohoo, 7 September 2022,
https://www.youtube.com/
watch?v=iagYicEvcfU.

33 Kourtney Kardashian.
"Photograph of Hawaiian Beach
by Author." *Instagram*, 23 April
2022, accessed 1 February
2023, https://www.instagram.
com/p/CcqLGzwFgZS/?utm_
source=ig_embed&ig_
rid=1836d37e-7a7c-4fd8-9746-
d818fad28306.

Chapter 18: You ... Yes, You

1 Drake Baer, "Dwight
Eisenhower Nailed a Major
Insight About Productivity,"
Business Insider, 11 April 2014,
https://www.businessinsider.
com/ dwight-eisenhower-
nailed-a-major-insight-about-
productivity-2014-4.

2 Roger Fisher, William L.
Ury, and Bruce Patton. 1991.
"Getting to Yes: Negotiating
Agreement Without Giving In."
London. Penguin Books.

3 Philippa Nuttall, "Don't Listen
To The Climate Doomists." *The
New Statesman*, August 2022,
https://www.newstatesman.
com/environment/
climate/2022/08/climate-
doomism-dont-listen-toxic-
narrative.

4 Marco Silva, "Why Is Climate
'Doomism' Going Viral –
And Who's Fighting It?" *BBC
News*, 23 May 2022, https://
www.bbc.com/news/blogs-
trending-61495035.

5   Justine Calma, "Climate
     Change Denial Is Making
     A 'Stark Comeback' On
     Social Media, Study Finds."
     The Verge, 20 January 2023,
     https://www.theverge.
     com/2023/1/19/23562269/
     climate-change-denial-
     social-media-meta-facebook-
     instagram-twitter.

6   Kari Paul, "Climate
     Misinformation on Facebook
     'Increasing Substantially',
     Study Says." *The Guardian,*
     5 November 2021, https://
     www.theguardian.com/
     technology/2021/nov/04/
     climate-misinformation-
     on-facebook-increasing-
     substantially-study-says.

7   Ibid., Nuttall.

8   Owen Mulhernaug, "The Anti-
     Plastic-Straw Phenomenon."
     Earth.org, 24 August 2020,
     https://earth.org/data_
     visualization/the-anti-plastic-
     straw-phenomenon/.

9   Seth Wynes and Kimberly
     A. Nicholas, "The Climate
     Mitigation Gap: Education and
     Government Recommendations
     Miss the Most Effective
     Individual Actions," *IOP
     Publishing* (2017), https://doi.
     org/10.1088/1748-9326/aa7541.

10   James Clear. 2018. "Atomic
     Habits." London. Penguin
     Random House.

The Author

John Pabon has spent two decades in the business of saving our Earth. After leaving his role at the United Nations, he travelled the world studying the impacts of sustainability first-hand in factories, on fields, and at Fortune 500s.

John is a globally recognised expert in sustainability. His career has taken him from Los Angeles to New York, Shanghai to Seoul to Melbourne. He has worked with the United Nations, McKinsey, A.C. Nielsen, and as a consultant with BSR, the world's largest sustainability-focused business network. A decade of experience living and working in Asia inspired him to found strategic communications firm Fulcrum Strategic Advisors, with a mission to help companies, governments, and individuals capitalise on the benefits of sustainability.

John is a regular contributor to major publications such as CNN, EuroNews, and the ABC. He also speaks to an array of global audiences on issues of sustainability, geopolitics, communications, and societal change. He is the Chair of The Conference Board's Asia Sustainability Leaders Council, a member of the United Nations Association of Australia, and serves on the board of advisors to the U.S. Green Chamber of Commerce.

John is the author of *Sustainability for the Rest of Us: Your No-Bullshit, Five-Point Plan for Saving the Planet*. Originally from Southern California, John currently lives in Melbourne, Australia, with his partner and their fussy Shiba Inu.